High Throughput Analysis
for Early Drug Discovery

High Throughput Analysis
for Early Drug Discovery

Edited by

James N. Kyranos
ArQule, Inc.
Woburn, MA, USA

2004

ELSEVIER
ACADEMIC
PRESS

Amsterdam – Boston – Heidelberg – London – New York – Oxford – Paris
San Diego – San Francisco – Singapore – Sydney – Tokyo

ELSEVIER B.V.
Sara Burgerhartstraat 25
P.O. Box 211, 1000 AE
Amsterdam, The Netherlands

ELSEVIER Inc.
525 B Street, Suite 1900
San Diego, CA 92101-4495
USA

ELSEVIER Ltd
The Boulevard, Langford Lane
Kidlington, Oxford OX5 1GB
UK

ELSEVIER Ltd
84 Theobalds Road
London WC1X 8RR
UK

First edition 2004

Library of Congress Cataloging in Publication Data
A catalog record is available from the Library of Congress.

British Library Cataloguing in Publication Data
A catalogue record is available from the British Library.

ISBN: 0-12-431165-2

⊗ The paper used in this publication meets the requirements of ANSI/NISO Z39.48-1992 (Permanence of Paper).
Printed in The Netherlands.

Working together to grow libraries in developing countries

www.elsevier.com | www.bookaid.org | www.sabre.org

ELSEVIER   BOOK AID International   Sabre Foundation

Contents

Preface

During the last two decades, the pharmaceutical industry has been under enormous and ever-increasing pressure from a variety of interest groups to enhance the productivity and effectiveness of drug discovery and development. Although, the collective R&D budget for the industry has been increasing exponentially for the past twenty years, the numbers of new chemical entities that reach the market have remained relatively constant during this same time period. Much of the increased investment has been in a variety of technologies focused on enhancing early drug discovery, such as high throughput screening, combinatorial chemistry, parallel synthesis, genomics, and automation to name just a few. Combinatorial and high throughput parallel synthesis chemistry have been the most recent potential levers developed to bridge the early discovery productivity gap. Although initial debate focused on the merits and limitations of split-and-mix vs. spatially addressable arrays or solid vs. solution phase approaches, over the last few years the emphasis has shifted to purification and characterization support as an integral part of the synthesis process.

The traditional medicinal chemistry approach of making one to ten compounds in parallel with significant quantities never challenged the available analytical methods for analysis and characterization. A typical round bottom flask synthesis produced enough material for NMR analysis, which is the preferred characterization technique of the synthetic chemist. However, with the introduction of mix-and-split and spatially addressable formats that now routinely generated 100 to 10,000 analogs from a scaffold, the analytical approach became a significant issue and the traditional medicinal chemistry approach of purifying and characterizing each compound synthesized was challenged because the tried and true analysis techniques like NMR could not initially be applied in a high throughput mode. The overriding need to correlate biological activity with molecular structure either before or after screening and use the information to follow up with synthetic modifications to parts of the molecule that appear to regulate potency, selectivity or any other parameter that needs to be optimized has driven the industry to identify, develop and implement analytical solutions that are appropriate and adaptable to the synthesis approach.

The need to adapt and develop analytical techniques and methods in response to synthetic constraints can be compared to the typical evolutionary process that selects for the most appropriate attributes in species selection. The attributes or methods that solve the problem in question will continue to be valued, honed and developed into highly efficient and specialized techniques. Two analytical techniques that were identified early in the development of high throughput organic chemistry and have continued to play an important role are high performance liquid chromatography (HPLC) and mass spectrometry (MS). These techniques alone or more importantly in combination have continued to generate the most interest. Much of the appeal of these techniques is the inherent high sensitivity, which translates into small quantity requirements, low volumes of injection, confirmation of molecular structure in the case of MS and a tradition with significant experience in automated unattended runs, which addresses the high throughput need. Moreover, the flexibility and adaptability of chromatography and mass spectrometry have contributed to the preferential selection of these techniques to support high throughput organic synthesis regardless of the various permutations that have been developed and successfully implemented.

Although the general analytical requirements of sensitivity, confirmation of structure and automated unattended operation are shared by the various high throughput organic synthesis techniques, the final embodiment of these parameters is ultimately determined by the synthesis methodology and the final form of the compounds of interest. For instance, if one were to synthesize libraries on solid supports using mix-and-split techniques, the final analytical methodology would have to accommodate single bead analysis with possibly only tens to hundreds of nanograms of total quantity. On the other hand, if one were to use solution phase, spatially addressable parallel synthesis techniques that generate tens to hundreds of milligrams of total quantity, then different embodiments of chromatography and mass spectrometry may be more relevant and appropriate.

The aim of this book is to provide the reader with an understanding of the various analytical methodologies that have been successfully developed and implemented in a variety of industrial settings in response to the various organizational strategies of high throughput organic synthesis, and arm the reader with an arsenal of options that can be adapted to one's individual synthetic needs and requirements. The book begins with one of the earliest approaches of combinatorial chemistry using mix-and-split methodologies and reviews the analytical approach of encoding and decoding individual beads. The initial analytical challenge of the completely randomized mix-and-split approach that generated mixtures of large numbers of compounds favored the development of a modified split-and-mix method using discrete containers that can be followed through the process by a variety of mechanical and electronic identifiers. This has culminated in the Irori MicroKan system and the associated analytical characterization method presented in Chapter 2. Flow injection analysis using mass spectrometry (FIA/MS) was one of the earliest implementations of MS in this field since it provides direct confirmation of the anticipated molecular structure and has continuously been modified and adapted to provide quality control support of syntheses from a variety of approaches. Chapters 3 and 4 provide insight into the use of FIA/MS to solve particular challenges requiring rapid analysis of very large sample numbers, as well as the most recent approaches to multiplexing different analytical streams using one mass spectrometer, thus saving time, resources and lab space. Chapter 5 presents the analytical challenges of using solution phase spatially addressable chemistry to synthesize large numbers of compounds in relatively large quantities and the approach of characterizing and quantifying each compound using rapid HPLC/MS, thereby industrializing the traditional medicinal chemistry approach. Chapter 6 presents a similar analytical characterization approach for the analysis of spatially addressable libraries even though the synthesis method leverages solid phase techniques but also leverages the multiplexing capabilities of new HPLC/MS instrumentation in a pseudo parallel approach. One of the earliest approaches to industrializing high throughput purification and the characterization needs for fraction identification and final compound characterization using and FIA/MS and HPLC/MS, respectively, is incorporated in Chapter 7. HPLC/MS has clearly become the dominant analytical technique to support high throughput synthesis and electrospray and atmospheric pressure chemical ionization are the two most common techniques associated with current HPLC/MS analysis. However, a new approach of using capillary HPLC coupled to MALDI-TOF/MS is presented in Chapter 8, which re-casts the traditional role of mass spectrometry in a new light. Finally, in a similar fashion Chapter 9 re-evaluates the role of NMR for characterization of compounds developed by high throughput synthesis, given the significant technological improvements that have been made in the past ten to fifteen years.

In addition to the techniques and methodologies presented here, there are other techniques and/or variants to these that have been investigated, developed and implemented for specific applications. The ones that have successfully addressed important questions raised by the synthesis process have survived and in many cases thrived. As the high throughput organic synthesis approaches continue to develop and move in a variety of different directions in response to the needs and pressures of modern drug discovery, the evolutionary pressure on analytical chemistry will also continue to force change, adaptation and innovation in order continuously to find effective solutions to the new challenges.

James N. Kyranos

List of Contributors

Lynn M. Cameron, Ph. D.
Senior Field Applications Specialist
Applied Biosystems
850 Lincoln Centre Drive
Foster City, CA 94404
Email: camerolm@appliedbiosystems.com

Cheryl D. Garr, Ph.D.
Director, Business Development
Albany Molecular Research, Inc.
18804 North Creek Parkway
Bothell, WA 98011
Email: cheryl.garr@albmolecular.com

Peter W. Davis, Ph.D.
Nereus Pharmaceuticals
10480 Wateridge Circle
San Diego, CA 92121
Email: pwdavis@JL-Sciedit.com

Samuel W. Gerritz, Ph.D.
Bristol-Myers Squibb
5 Research Parkway
Wallingford, CT 06492
Email: samuel.gerritz@bms.com

Jeannine Delaney, Ph.D.
Pfizer Discovery Tech Center
620 Memorial Drive
Cambridge, MA 02139
Email: Jeannine_Delaney@cambridge.
pfizer.com

Michael C. Griffith, Ph.D.
University of California, San Diego
UCSD Extension, Dept. 0170-M
9500 Gilman Dr.
La Jolla, CA 92093
Email: ucxgrif@san.rr.com

Lawrence W. Dillard
Pharmacopeia, Inc.
PO Box 5350
Princeton, NJ 08543-5350

Joan Guo
Pharmacopeia, Inc.
PO Box 5350
Princeton, NJ 08543-5350

Roland E. Dolle
Adolor Corporation
700 Pennsylvania Drive
Exton, PA 19341

Ian Henderson, Ph.D.
Pharmacopeia, Inc.
PO Box 5350
Princeton, NJ 08543-5350
Email: ian@pharmacop.com

Nelson Huang
Wyeth Research
Chemical & Screening Sciences
200 Cambridge Park Drive
Cambridge, MA 02140
Tel: 617-665-5632

Lesline V. Julien
ArQule, Inc.
19 Presidential Way
Woburn, MA 01801
Email: ljulien@arqule.com

Oliver Keil
Merck Frankfurter Str. 250
D-64293 Darmstadt
Germany

James N. Kyranos, Ph.D.
ArQule, Inc.
19 Presidential Way,
Woburn, MA 01801
Email: jkyranos@arqule.com

Heewon Lee, Ph.D.
Chemical Development Department
Boehringer Ingelheim
Pharmaceuticals, Inc.
900 Ridgebury Road
Ridgefield, CT 06877
Tel: 203-791-6736
Email: hlee@rdg.boehringer-ingelheim.
com

Tammy LeRiche
Merck
Frankfurter Str. 250
D-64293 Darmstadt
Germany

Kenneth L. Morand, Ph.D.
Procter & Gamble Pharmaceuticals
Health Care Research Center
8700 Mason-Montgomery Road
Mason, OH 45040
Email: morand.kl@pg.com

Christine Salvatore
SRI International
4111 Broad Street
San Luis Obispo, CA 93401-7903
Email: r.salvatore@comcast.net

Christopher R. Sarko, Ph.D.
Medicinal Chemistry Department
Boehringer Ingelheim
Pharmaceuticals, Inc.
900 Ridgebury Road
Ridgefield, CT 06877
Tel: 203-791-6147
Email: csarko@rdg.boehringer-ingelheim.
com

Paul D. Schnier
Amgen
Discovery Analytical Sciences
One Amgen Center Dr.
Thousand Oaks, CA 91320
Tel: 805-447-4302
Email: pschnier@amgen.com

Lauri Schultz, Ph.D.
Scientist II
CEPTYR, Inc.
3830 Monte Villa Parkway
Bothell, WA 98021

Andrea M. Sefler
Drug Metabolism and Pharmacokinetics
GlaxoSmithKline
5 Moore Drive
RTP, NC 27709
Email: andrea.m.sefler@gsk.com

Mary M. Sherman, Ph.D.
Director, ADME
MPI Research, 54943 North Main Street
Mattawan, MI 49071-9339
Tel: 269-668-3336x1480

Marshall M. Siegel, Ph.D.
Wyeth Research
401 N. Middletown Rd.
Bldg 222/Room 1043
Pearl River
NY 10965
Email: SIEGELM@wyeth.com

Hui Tong
Wyeth Research Chemical Sciences
401 N. Middletown Rd.
Building 222/Room 1044
Pearl River, NY 10965
Tel: 845-602-2628

Craig S. Truebenbach
Wyeth Research
DSM-Biotransformation
500 Arcola Rd, S3416
Collegeville PA 19525
Tel: 484-865-3073

Dietrich A. Volmer, Ph.D.
Institute for Marine Biosciences
National Research Council
1411 Oxford Street Halifax,
Nova Scotia B3H 3Z1
Canada
Email: Dietrich.Volmer@nrc.ca

Acknowledgments

This book has been an ongoing project for quite some time now and has evolved to its present state by the sustained efforts of several original authors and the willing contribution of new ones who joined later. I want to thank all my colleagues who participated on this project for their enthusiasm in agreeing to contribute their work and for maintaining their commitment to see this project become a reality.

I would also like to thank Dr. Hong Cai for her efforts in helping coordinate the project early in the book's development and especially acknowledge Michele Livingston, my assistant, for coordinating many of the technical and communication activities that pulled all the chapters, figures, captions and references in a consistent format ready for publication.

Finally, I would like to thank my wife and children for their patience, understanding and support during many weekends when much of my time was focused on this project and would like to dedicate this book to them as a small token of my appreciation of them.

1

High Throughput Analysis of Combinatorial Libraries Encoded with Electrophoric Molecular Tags

Ian Henderson, Joan Guo, Lawrence W. Dillard and
Mary M. Sherman

Pharmacopeia, Inc., P.O. Box 5350, Princeton, NJ 08543-5350, USA

Roland E. Dolle

Adolor Corporation, 700, Pennsylvania Drive, Exton, PA 19341, USA

CONTENTS

1 INTRODUCTION

Encoded combinatorial libraries of small molecules are a valuable resource for the discovery of biologically active agents.[1-3] Originally conceived to simplify the process of deconvolution and compound identification in split-and-pool libraries,[4-6]

High Throughput Analysis for Early Drug Discovery
Edited by James N. Kyranos

a variety of strategies have now been described to encode libraries.[7-35] One strategy, successfully pioneered at Pharmacopeia, is a binary encoding protocol employing electrophoric molecular tags (ECLiPS™ technology).[7-10,36-47] In this protocol, sets of synthons are serially combined through split-and-pool or direct divide[48] synthesis in tandem with the incorporation of binary sets of electrophoric tags on solid support. Each bead in an encoded library contains a compound, whose synthesis history is recorded by a unique set of attached tags. Because orthogonal reactive linkers are used in the construction of the library, compound and associated tags can be independently released allowing for off-bead assays. The identity of any given library member is readily inferred through the process of decoding, i.e. gas chromatography/electron capture detection (GC/ECD) analysis of detached tags.

The characterization of large encoded combinatorial libraries created through split-and-pool synthesis is a challenging venture.[1-6] In addition to the concerns of analysis speed, data capture, and universal quantitation, the library size and per bead yield adds another dimension of complexity. A 50 000-member library encoded with electrophoric molecular tags, prepared in 200-fold redundancy contains 10 million beads. Depending on the resin (bead) loading and the overall yield of the synthetic sequence, each bead in the library possesses ≤ 1 nmol of compound and 1–5 pmol of associated tags. The enormity of a library's bead count and the sub-nanomole quantity of compound present on each bead preclude characterization of every compound. A more appropriate and pragmatic approach is to develop, validate, and implement quality control protocols that ensure a high degree of library fidelity.

While extensive synthon profiling, careful reaction optimization, rigorous analysis of library QC compounds are necessary to complete a successful library synthesis, it is still highly desirable to have a much broader knowledge of the chemistry that occurs during actual library construction, and assurance that the compounds eluted from the beads are physically present in the wells of assay plates.[49-56] There are six quality control measures routinely employed at Pharmacopeia for library creation[47,57-59] and compound elution (Fig. 1.1). These include: (1) QC compound synthesis, (2) tag QC, (3) construction-in-process QC, (4) intermediate quality assurance (QA), (5) final library QA, and (6) production-in-process QC.[59] Quality control procedures (1)–(3) occur during the library synthesis cycle. Production-in-process QC is a protocol deployed in the production department to ensure the proper elution of compounds from beads into wells of assay plates on a mass scale. Intermediate and final library QA are protocols that provide conformation on the overall success of the synthesis and the performance of individual synthons.

With the primary goal of assessing the overall quality of an encoded combinatorial library, a statistical sampling procedure was devised which combines the application of single bead LC/MS analysis with tag decode analysis to confirm the existence of putative library compounds.[60] The statistical-based, tag decode-assisted single bead LC/MS analysis, is conveniently termed "library quality assurance" (library QA). The development and validation of library QA as a useful

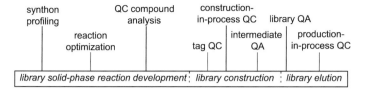

Synthon profiling/reaction optimization: Confirmation of reagent viability, establishing suitability of monomer combinations in the proposed synthesis. >50% profiling of one monomer set against representative members of a second set. Gravimetric yield and purity analysis obtained for >30 compounds.

QC compound analysis: 5–10 library compounds synthesized as an-bead and purfied off-bead standards rigorously analyzed to estimate overall library yield and purity.

Tag QC: Tag fidelity confirmed after each tagging step prior to pooling or direct divide.

Construction-in-process QC: Resynthesis of 2–3 QC compounds during actual library construction using bulk reagents and solvents.

Intermediate library QA: Decode-assisted single-bead LC/MS to confirm physical existence of cleaved library members. Provides information regarding composite library fidelity and the performance of individual synthons.

Production-in-process QC: Production safeguard to ensure optimal elution of library compounds into essay plates.

Figure 1.1   QC protocols along the library pipeline.

qualitative measure of library fidelity is demonstrated here in the evaluation of a 25 200-member library of statine amides.

2   TAG DECODE-ASSISTED SINGLE BEAD LC/MS ANALYSIS (LIBRARY QA)

Libraries prepared by parallel synthesis typically provide a manageable number of compounds (<5000) in sufficient quantity (>0.1 mg) to allow identity, purity, and yield for all members (or a statistically relevant number) to be determined.[61-64] This may be accomplished by employing routine HPLC analysis coupled singularly or in combination with UV,[65-70] evaporative light scattering detection (ELSD),[71-75] chemiluminescent nitrogen detection (CLND),[76-78] MS,[79-89] detection and more recently, high throughput NMR analysis.[90] Automated HPLC may also be used to optionally purify every compound in a library prior to biological testing.[91,92] These approaches to library characterization and compound purification are generally not feasible with encoded combinatorial libraries, where 200–300 copies of a library (*e.g.* 10–15 million beads per 50 000 member library), are routinely produced. The average amount of compound present on one bead is approximately 200 pmol, hence, high throughput compound analysis is, in essence, restricted to the more sensitive analytical technique of LC/MS.

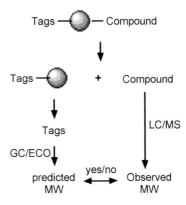

Figure 1.2 Single bead library
QA process.

One approach to characterizing the large encoded libraries is through library QA analysis. By combining LC/MS analysis with tag decode analysis, the physical existence of the structure of any given library compound can be corroborated. Figure 1.2 illustrates the library QA process. The compound from a single bead is eluted under conditions optimized for its particular library. The tags from the same bead are oxidatively removed and analyzed by GC/ECD, and a predicted molecular weight for the decoded compound generated. The compound (bead eluent) is then analyzed by LC/MS, to determine its molecular weight. Comparing the molecular weight predicted by the tag decode with the empirical value yields a "yes/no" answer corroborating the presence or absence of the inferred compound. Statistical sampling of several hundred randomly selected beads from a library provides information on library fidelity and an indication of the performance of individual library synthons.

3 STATISTICAL CONSIDERATIONS

The goals of sampling an encoded combinatorial library are to determine the overall success of the solid-phase synthesis and to assess the performance of each synthon. Tag decodes reveal, which compounds (synthons) are supposed to be on each bead, and LC/MS analysis reveals whether they are physically present. It is impractical to analyze all the compounds in an encoded library where the total bead count is in excess of one million; therefore, statistical sampling techniques are required.[93-95] The application of this approach to Pharmacopeia libraries has been previously reviewed.[58]

Library QA provides a very good estimate of the quality of a combinatorial library on two levels. First, a highly precise measure of the proportion of compounds successfully synthesized in a library is calculated. The precision in the estimate of library fidelity is high because the sample size is on the order of several hundreds. Library QA permits the scientist to identify and readily distinguish between poor, mediocre, and good quality combinatorial libraries. Second, statistical analysis of the sampling data can assist the chemist in ascertaining the performance of *individual* synthons in a library, with the proviso that the synthon sample size, $n_{i,j}$, must be sufficiently large (> 10).[60] Library QA provides no information regarding the *combinatorial* synthesis success of synthons or intermediates, as the required n would exceed a practical limit. Rather, the analysis provides information on average proportions of observed compounds and synthons in a library.

4   SYNTHESIS OF THE ENCODED STATINE AMIDE LIBRARY

The encoded solid-phase synthesis of the statine amide library **1** (Scheme 1.1 and Fig. 1.3), has been previously described.[58] Statine library **1**, synthesized using two combinatorial steps and a two-step N-derivatization sequence, is

Scheme 1. Synthesis of the statine library **1**.

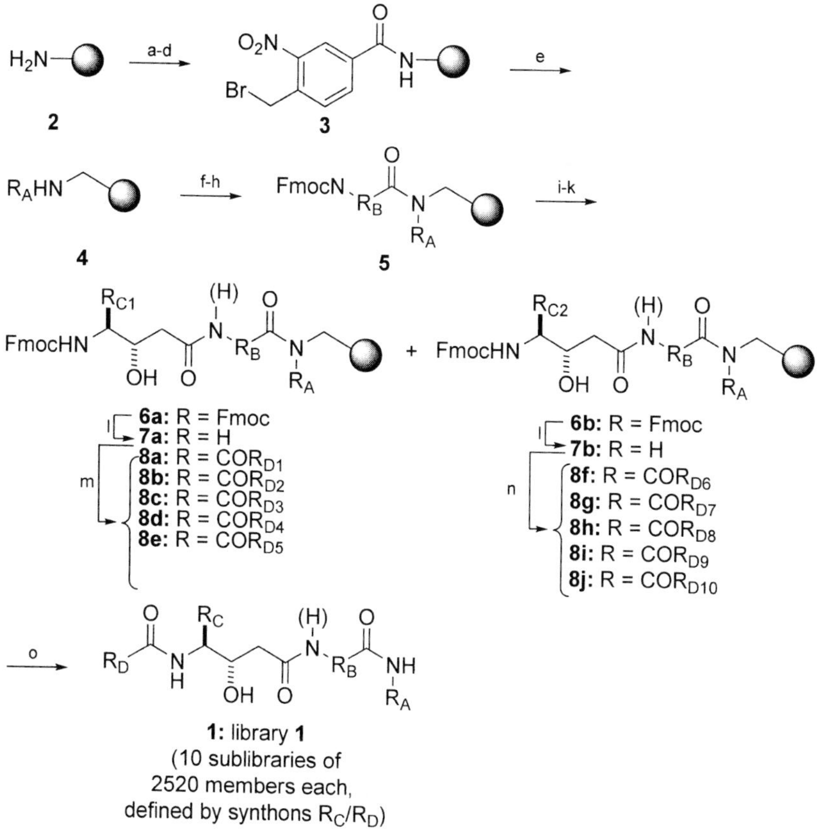

Scheme 1.1   Synthesis of statin library **1**.

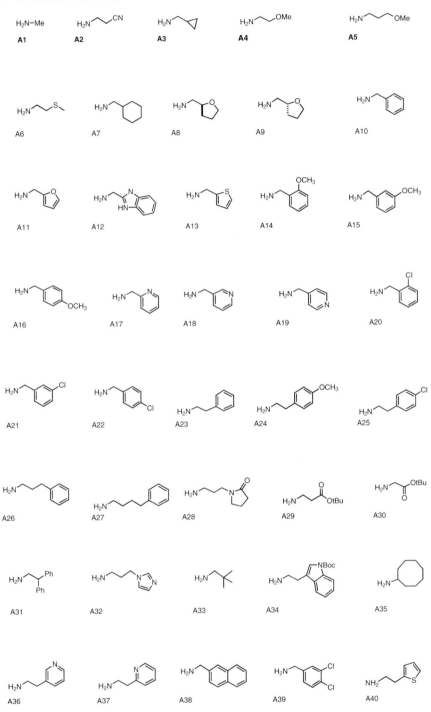

Figure 1.3   Synthons used for statin library **1**.

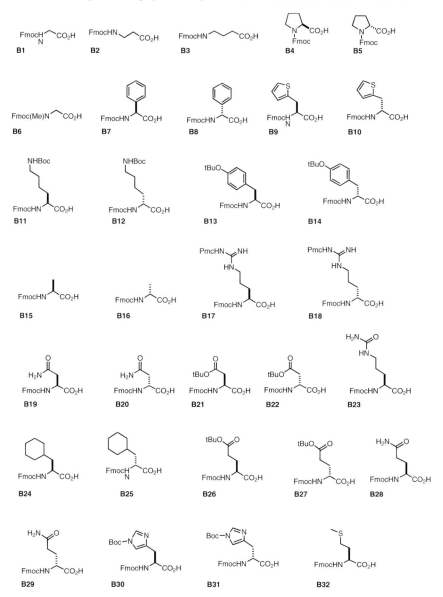

Figure 1.3 *(continued).*

composed of 10 sub-libraries with 2520 unique compounds per sub-library
(: $40(R_A) \times 63(R_B) \times 10(R_C/R_D) = 2520$). To assess the precision of the library
QA protocol, analysis of this library was performed in triplicate. The target sample
size n for the QA analysis was 630 beads, corresponding to 10× the largest synthon
set – **R_B**. In practice, for each run 60–65 beads per sub-library were arrayed as
single bead per well in 96-well filter bottom plates. The beads were suspended in a

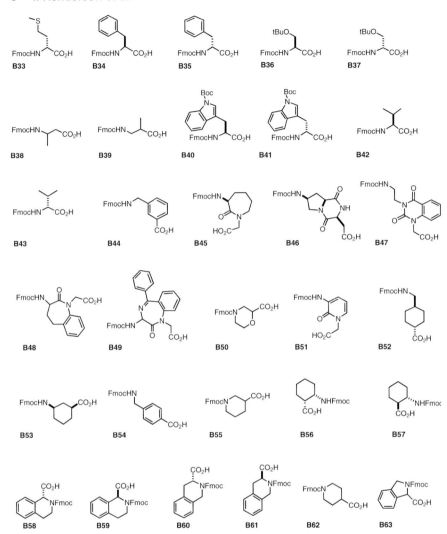

Figure 1.3 (*continued*).

solution of aqueous ethanol containing 3% TFA and irradiated at 365 nm for 30 min according to a previously determined elution protocol.[57–59] Bead eluents, containing the detached compounds, were filtered into derivative plates and dried *in vacuo*. The beads were decoded, generating a structure and a predicted molecular weight for each compound. The dried compounds in the derivative plates were redissolved in a small volume of acetonitrile–water (4:1) and analyzed by LC/MS. Mass spectral analysis was carried out in the positive ion mode only. Due to the small quantities of cleared compound obtained from the beads and the interference of bead matrix present in reconstituted samples, total ion chromotograms (TIC)

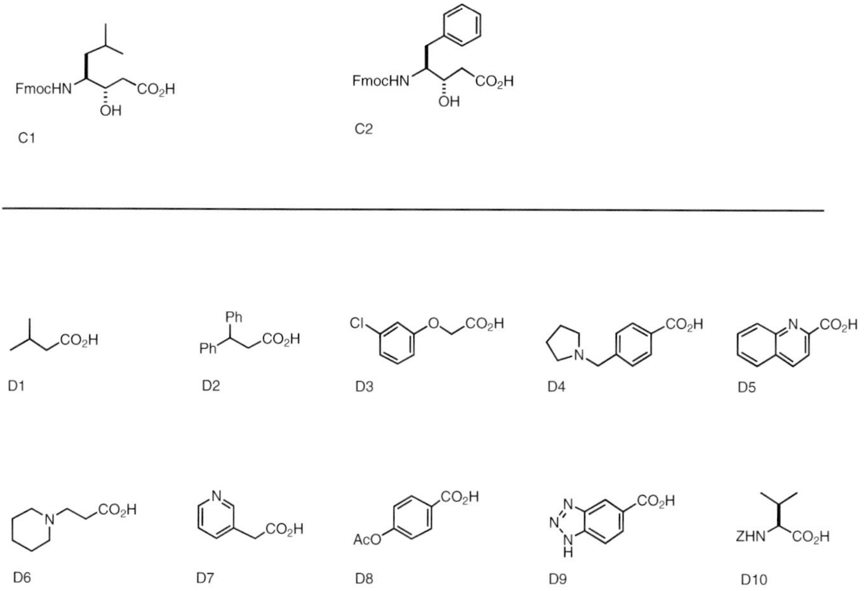

C1

C2

D1   D2   D3   D4   D5

D6   D7   D8   D9   D10

Figure 1.3 (*continued*).

could not be used to confirm putative library compounds. The use of extracted ion-chromatograms (XIC), which is a standard practice in drug metabolism where biological matrix may obscure trace compounds of interest was used for putative compound confirmation. If the XIC chromatogram revealed the expected mass $(M + H)^+$ above a 3:1 signal to noise ratio, then a "found (F)" answer for that library bead was secured. If the expected mass ion was not detected, or below the threshold criteria then a "not found (NF)" was recorded. In approximately 2–3% of the "found" answers, the S/N ratio was $>3:1$ but the peak height was in the lower end of the detector response $(<4.0 \times 10^4)$. A custom data analysis program automatically performed the molecular weight comparisons.

Statistical data are presented for all three composite analyses (Fig. 1.4a). Comparative answers are colored coded: found (F; gray) and not found (NF; black). The proportion p of confirmed compounds was computed by dividing the number of compounds found by the total number of beads analyzed.

The sample size n for the first run (QA run-1) was 610 beads (Fig. 1.4a, Table 1.1). The total number of found answers F $((M + H)^+)$ was 515 giving rise to a found proportion, $p = 84.4\%$. For the two additional analyses conducted, the found proportions are $p = 80.2$ and $p = 84.7\%$, respectively. Thus the three replicate QA analyses are in excellent agreement with one another, consistent with statistical expectations.

Across the three QA runs, a total of 1902 beads was decoded and analyzed by single bead LC/MS, representing a library sampling of 7.6%. Given that the library was prepared in 200-fold redundancy, a number of replicate structures are expected

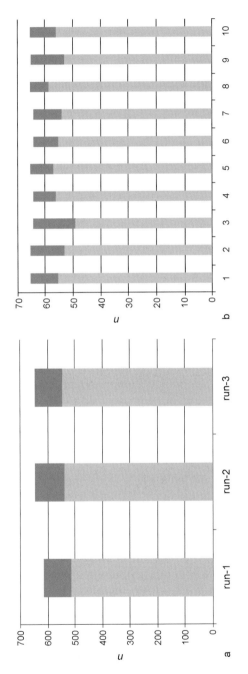

Figure 1.4 (a) Composite library **1** QA analysis and (b) sub-library (**C/D** synthon) analysis for QA run-3.

Table 1.1 Composite and sub-library QA analysis for library 1

QA	Run 1				Run 2				Run 3			
	n	F	p (%)	lb–ub (%)	n	F	p (%)	lb–ub (%)	n	F	p (%)	lb–ub (%)
Composite sub-library	610	515	84.4	81.5–87.2	645	536	83.1	80.2–85.9	646	547	84.7	81.9–87.5
1	61	53	86.9	77.1–95.5	65	54	83.1	72.8–92.4	65	55	84.6	74.6–93.6
2	63	56	88.9	79.8–96.8	63	53	84.1	73.9–93.3	65	53	91.2	81.5–95.0
3	61	52	85.2	75.1–94.3	65	53	81.5	70.9–91.2	64	49	76.6	65.1–87.3
4	62	49	79.0	67.7–89.5	65	51	78.5	67.3–88.7	64	56	87.5	78.1–95.7
5	62	56	90.3	81.6–97.8	65	52	80.0	69.1–90.0	65	57	87.7	78.5–95.8
6	62	47	75.8	64.8–86.8	64	50	78.1	66.9–80.5	64	55	85.9	76.2–94.6
7	63	48	76.2	64.5–87.0	65	53	81.5	70.9–91.2	64	54	84.4	74.3–93.5
8	57	50	87.7	77.8–96.4	65	55	84.6	74.6–93.6	65	59	90.8	82.5–97.9
9	60	51	85.0	74.7–94.2	64	56	87.5	78.1–95.7	65	53	91.2	81.5–95.0
10	59	53	89.8	80.7–97.6	64	59	92.2	84.3–98.8	65	56	86.2	76.5–94.7

during decoding. Experimentally, 65 replicate structures were found. The F vs NF assignments agreed for 61 of the replicates or 94% (data not shown). The high reproducibility of the comparative assignments observed for the replicate structures, together with the consistent composite p values of the three QA analyses, provide a satisfactory level of validation for this statistical-based approach to library quality assessment.

Examination of the statistical results obtained for each sub-library reveals that the F and NF answers are fairly evenly distributed (Table 1.1). The proportion of positively identified compounds in QA run-3 ranged from 76.6% (sub-library **3**) to 91.2% (sub-libraries 2 and 9). The data implies that there were no gross synthetic failures in the two-step derivatization sequence in which the C/D synthons were coupled to resin-bound intermediate **5** (Scheme 1.1).

Statistical information regarding the performance of the **A** synthons may also be obtained from the library QA analysis (Fig. 1.5 and Table 1.2). Recorded p_i values range from 33.3% (**A32**, QA run-3) to 100% (*e.g.* **A3**; QA run-3) with the majority (*ca.* 35 of 40) of the **A** synthons having p_is $> 75\%$. As a group, the hydrophobic synthons, including the aliphatic amines (*e.g.* **A1**, **A3**, **A7**), arylalkyamines (*e.g.* **A10**, **A15**, **A20**), and the amino ethers (*e.g.* **A4**, **A5**, **A8**, **A9**), were strong performers, suggesting that they were successfully incorporated into the library. Noteworthy, are the relatively high average proportions ($p_i > 80\%$) found for the hindered amines, neopentylmethylamine (**A33**) and cyclooctylamine (**A35**).

One exception to this trend is the neutral, hydrophobic synthon, methylthioethylamine (**A6**). The proportion $p_{A6} = 50\%$, in QA run-3, and comparable p_{A6} values found in the other two QA runs: $p_{A6} = 55.6\%$ (QA run-1), and $p_{A6} = 41.7\%$ (QA run-2) indicate that the synthon's performance is mediocre, which is likely a reflection of its propensity to undergo photo-oxidation upon release from the bead. In all cases where the A6-contained compounds were NF in the QA analysis, the corresponding sulfoxide $(M + OH)^+$ was found as a major product in the respective XIC chromatograms. It is reasonable to assume that the addition of amine **A6** to resin (step **3** to **4**, Scheme 1.1) occurred in high yield, analogous to other structurally related, successfully incorporated amines, *e.g.* synthon **A4** (p_{A4}, 88.9%). It can be argued too, that the poor performance of furfurylamine (**A11**; $p_{A11} = 57.1\%$) may also be due to the known photo-oxidation of this heterocycle.

In contrast to the performance of the neutral, hydrophobic synthons, the performance of hydrophilic synthons possessing positive or negative charged atoms or functional groups is more variable. For example, the proportions determined for the pyridinylalkylamines (**A17**, **A18**, **A19**, **A36** and **A37**) are $p \sim 50\%$. The 3-aminopropylimidazole (**A32**) synthon is an especially poor performer: $p_{A32} = 33.3\%$. On the basis of QA results, it is likely that many of the putative compounds in the library possessing these synthons are absent. It is unclear whether the divergence between the performances of hydrophobic and hydrophilic **A** synthons, is a result of the chemical failure, the poor diffusion of these compounds out of the bead matrix, photochemical sensitivity, or a combination of these factors. Another possibility for the tendency to see higher overall performance of the hydrophobic synthons and a higher confirmation rate of lipophilic compounds may be in part due

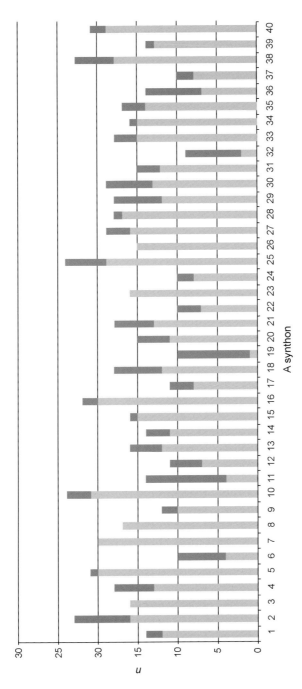

Figure 1.5   A synthon analysis (QA run-3).

Table 1.2   Proportion (p) and boundaries (lb–ub) for selected A synthons

QA	Run 1				Run 2				Run 3			
	n	F	$p(\%)$	$lb-ub$ (%)	n	F	$p(\%)$	$lb-ub$ (%)	n	F	$p(\%)$	$lb-ub$ (%)
Synthon A												
1	16	14	87.5	69.8–100	13	9	69.2	39.4–95.9	14	12	85.7	65.7–100
3	17	16	94.1	80.8–100	13	10	76.9	53.0–100	16	16	100	89.1–100
4	15	15	100	81.9–100	10	9	90.0	67.8–100	18	16	88.9	73.0–100
5	18	17	94.4	81.8–100	6	6	100	60.7–100	21	20	95.2	84.3–100
6	9	5	55.6	17.9–91.6	12	5	41.7	10.9–73.8	10	5	50.0	14.8–85.2
8	15	13	86.7	67.9–100	22	20	90.9	77.7–100	17	17	100	83.8–100
9	17	16	94.1	80.8–100	6	4	66.7	25.9–100	12	10	83.3	60.3–100
11	22	17	77.3	55.6–95.5	8	3	37.5	6.0–77.7	14	8	57.1	27.6–85.7
12	16	12	75.0	49.6–97.1	16	8	50.0	22.7–77.3	11	9	81.8	56.9–100
13	11	11	100	76.2–100	17	14	82.4	63.4–100	16	13	81.3	61.2–100
15	11	11	100	76.2–100	15	12	80.8	58.8–100	16	15	93.8	79.6–100
17	12	9	75.0	49.4–100	16	15	93.8	79.6–100	11	10	90.9	70.6–100
18	16	12	75.0	49.6–97.1	13	9	69.2	39.4–95.9	18	12	66.7	41.4–90.0
19	17	11	64.7	38.6–89.1	13	6	46.2	16.0–76.9	9	5	55.6	17.9–91.9
22	11	11	100	76.2–100	15	14	93.3	78.3–100	10	7	70.0	40.4–100
23	10	9	90.0	67.8–100	7	7	100	65.2–100	16	16	100	82.9–100
29	18	18	100	84.7–100	9	7	77.8	48.2–100	18	15	83.3	65.3–100
30	10	8	80.0	53.0–100	6	5	83.3	47.9–100	19	16	84.2	67.0–100
32	11	5	45.5	12.6–79.2	15	8	53.3	24.9–81.3	9	3	33.3	0.0–54.3
33	15	13	86.7	67.9–100	13	13	100	79.4–100	18	16	88.9	73.0–100

to an artifact of the LC/MS analysis. For example, LC is conducted employing a generic 5 min gradient on a C18 column (see experimental). Compounds having multiple charged groups, (*e.g.*, the putative compound **A8-B14-C1-D4**) may not be retained on the column and hence registered as NF.

Figure 1.6 displays the histogram for the F and NF answers for a set of **B** synthons (QA run-3). Sample sizes n_B range from 3 to 16 with the average $n_B = 10$ beads (647 beads/63 **B** synthons). As the ratio of the number of samples to the number of synthons narrows, the interpretation of the library QA data is only clear for very good or very poor performing synthons. In the case of QA run-3, step 2, there are a number of synthons that meet this criteria. Glycine (**B1**), sarcosine (**B6**), the acyclic β - and γ -amino acids (**B2**, **B3**, **B38**, **B39**), the hydrophobic α -amino acids including L- and D-prolines (**B4**, **B5**) and L- and D-phenylglycines (**B7**, **B8**), tetrahydroisoquinoline carboxylic acids (**B58-B61**), and many of the unnatural, peptidomimetic-type amino acids (**B45**, **B47-B57**, **B62**, **B63**), may all be regarded as well-performing synthons. The proportions observed for the asparagines (**B19**, **B20**), glutamines (**B28**, **B29**), L-citrulline (**B23**), and the methionines (**B32**, **B33**) are near $p = 50\%$; hence, with the narrow ratio of number of samples to number of synthons in the set, the true p for these synthons cannot be assessed with confidence. Other amino acids with hydrophilic side chains (aspartic acids, (**B21**, **B22**), glutamic acids (**B26**, **B27**), histamine (**B30**, **B31**), arginines (**B17**, **B18**), and the lysines (**B11**, **B12**) fared reasonably well, with p values close to 70. One synthon, **B46**, was clearly a problem in the library (Fig. 1.6 and Table 1.3). Proportion p_{B46} in QA run-3 equals 27, and for QA run-2, $p_{B46} = 30.8\%$. This synthon was not found at all in QA run-1: $p_{B46} = 0\%$. The poor performance synthon **B46** is believed to due to a chemical failure in the library. A statistical sampling QA analysis (>600 beads) was carried out on the tagged Fmoc-intermediate 5, and $F \sim 95\%$ was observed. There was sufficient quantity of material at this stage to obtain UV data for many of the intermediates, which showed $>85\%$ purity (unpublished observation). Specifically, synthon **B46** gave a $p > 80\%$ indicating chemical failure post coupling.

5 BIOLOGICAL SCREENING

Statines are a well-known class of transition-state isosteres displaying inhibitory activity against aspartic acid proteases.[96,97] Library **1**, was designed to explore the P_2 specificity of aspartic acid proteases and screened against two aspartic acid proteases, human cathepsin D (cat D),[98,99] and malarial plasmepsin II (plm II).[100] Solution-based assays for each enzyme were carried out using fluorescent energy transfer substrates. Screening was conducted *via* a two-part protocol. First, a survey screen was conducted in which 0.5 library equivalents (*ca.* 12 000 beads) were screened at *ca.* 30 beads per well to identify the most active sub-library. This was followed by screening two library equivalents of the most active sub-library at the single bead level (*ca.* 7500 beads).

Initially, 0.5 equiv. of library **1** was screened against cat D and plm II at *ca.* 30 compounds per well. The estimated screening concentration was 1 μM, based on

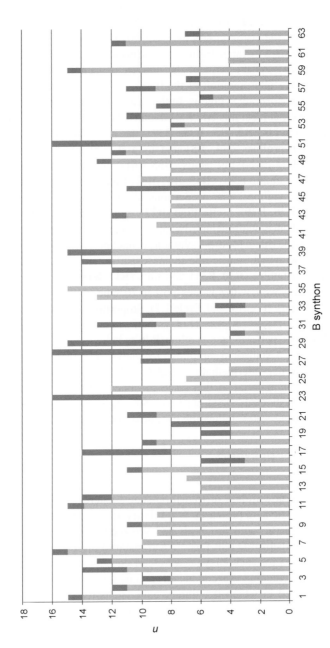

Figure 1.6   B synthon analysis (QA run-3).

Table 1.3 Proportion (p) and boundaries (lb–ub) for selection B synthons

QA	Run 1				Run 2				Run 3			
	n	F	p (%)	lb–ub (%)	n	F	p (%)	lb–ub (%)	n	F	p(%)	lb–ub (%)
Synthon B												
3	6	5	83.3	47.9–100	10	8	80.0	53.0–100	12	10	83.3	60.3–100
7	7	7	100	65.2–100	6	5	83.3	47.5–100	10	10	100	74.1–100
9	10	10	100	74.1–100	10	9	50.0	67.8–100	11	10	90.9	70.6–100
13	11	11	100	76.2–100	11	11	100	76.2–100	11	11	100	76.2–100
15	8	7	87.5	60.2–100	10	6	60.0	24.4–93.4	11	10	90.9	70.6–100
16	15	13	86.7	67.9–100	13	12	92.3	75.0–100	6	3	50.0	4.4–95.6
18	9	8	88.9	64.4–100	7	5	71.4	35.1–100	10	9	90.0	67.8–100
19	15	11	73.3	46.6–96.8	10	3	30.0	0.0–49.6	6	4	66.7	25.9–100
20	10	7	70.0	40.4–100	9	3	33.3	0.0–54.3	8	4	50.0	10.2–89.8
23	10	8	80.0	53.0–100	11	5	45.5	12.6–79.2	16	10	62.5	35.3–88.2
26	12	10	83.3	60.3–100	9	4	44.4	1.0–82.1	4	4	100	47.3–100
28	11	8	72.7	45.3–100	9	8	88.9	64.4–100	16	6	37.5	11.8–64.7
29	12	8	66.7	35.1–95.3	14	10	71.4	43.2–96.4	15	8	53.3	24.9–81.3
30	10	7	70.0	40.4–100	7	2	28.6	0.0–50.6	4	3	75.0	24.8–100
32	3	3	100	36.8–100	7	4	57.1	13.6–97.9	10	7	70.0	40.4–100
46	7	0	0.0	0.0–34.8	13	4	30.8	4.1–60.6	11	3	27.3	0.0–45.6
58	4	4	100	47.3–100	8	7	87.5	60.2–100	7	6	85.7	54.9–100
59	9	9	100	71.7–100	14	13	92.9	76.8–100	15	14	93.3	78.3–100
60	8	8	100	68.8–100	11	11	100	76.2–100	4	4	100	47.3–100
61	9	8	88.9	64.4–100	9	8	88.9	64.4–100	3	3	100	36.8–100

the bead yields of a set of QC compounds (data not shown). The highest activity was observed in sub-library **8j**, defined by the synthon pair **C2/D10**, followed by sub-library **8a**, defined by the synthon pair **C1/D1**. Subsequently, two additional follow-up copies of sub-library **8j** were screened one compound per well (5040 beads, 87% library coverage). Because of the exceptionally high potency observed in this sub-library during the survey screen, the estimated screening concentration was reduced to *ca.* 0.050 μM. Overall, sub-library **8j** appeared more active and selective against cat D than plm II.

Only those wells (single beads) that showed activity equal to or less than 30% of control activity remaining were decoded. This follow-up evaluation resulted in a total of 55 decoded structures for plm II, including 36 of which arose from three quadruplicate, two triplicate and nine duplicate structures, and 19 unique structures. For cat D, the follow-up evaluation resulted in 76 decoded structures, including 21 from two quadruplicate, 6 triplicate, and 14 duplicate structures, as well as 22 unique structures. The primary screening activity (% control remaining) obtained for the decoded, replicate plm II structures indicated marginal selectivity ($<$ 10-fold *vs* cat D). In contrast, the decoded replicate cat D structures showed selectivities ranging from 1:1 to $>$20:1.

Figure 1.7 provides a graphic representation of the R_A and R_B synthon frequencies for the decoded structures of sub-library **8j** against the two enzymes. Both enzymes preferred a rather broad range of R_A amines, providing these were neutral and hydrophobic. Plm II showed the largest range for its 32 decoded structures: 21 out of 55 **A** synthons were observed, including alkyl, cycloalkyl, and arylalkyl amines. For cat D, which was somewhat more selective, 15 out of 40 **A** synthons were observed for the 76 decoded structures; phenethylamine (**A23**; observed 12 times) had the highest frequency. Interestingly, the amines containing positive or negative charged groups, *e.g.* the pyridinylmethylamines (**A17–19, A36** and **A37**), glycine and the β-aminoalanine synthons (**A29–A30**) and imidazoly-lethylamine (**A32**) were not observed in the decoded structures, suggesting inhibitors with such substituents have estimated K_is $>$ 50 nM.

In contrast to the broad structural variation at position R_A, the enzymes were more stringent in their preferences for the R_B amino acids. Hydrophobic residues were again preferred over hydrophilic residues. L-Cyclohexylalanine (**B24**) and L-thienylalanine (**B9**) were found in 23 out of 32 decodes for plm II, along with L-tyrosine (**B13**), L-phenylalanine (**B34**) and L-valine (**B42**). Although cat D shared many of the same amino acids with plm II, including L-thienylalanine, L-tyrosine, L-phenylalanine, and L-valine, L-cyclohexylalanine **B24** was the most frequently observed. In addition, there were several **B** synthons that were unique to cat D, which appeared to impart a high degree of selectivity: D-tetrahydroisoquinoline-1-carboxylic acid (D-Tiq; **B58**), 3-amino-propionic acid (**B3**), 3-amino-cyclohexane carboxylic acid (**B53**), and the aminocaprolactam, (**B45**).

Five replicate structures were chosen for resynthesis. These included the two most potent plm II inhibitors (**9: A14-B24-C2-D10** and **10: A8-B24-C2-D10**, Table 1.4), and three of the most potent and selective cat D inhibitors (**11: A21-B45-C2-D10**; **12: A23-B53-C2-D10**; and **13: A11-B24-C2-D10**) as identified from the primary

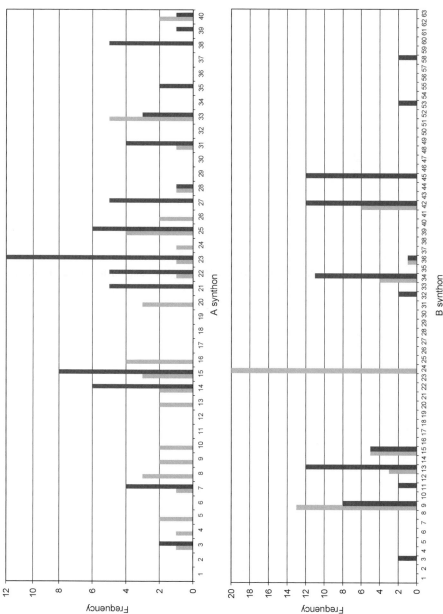

Figure 1.7 The **A** and **B** synthon frequencies for decoded structures from sub-library **8j** (plm II: black; cat D: gray).

Table 1.4   Inhibition constants (K_i) for resynthesized compounds

No.	Structure	K_i (nM)[a]	
		plm II	cat D
9[b]		29	44
10[c]		16	32
11[b]		210	3
18[d]		44,000	280
19[d]		18,000	290
20[d]		28,000	2,900
12[b]		29	4
13[b,e]		72	18
14[d]		12	44
15[d]		26	46
16[d]		15	140

(continued)

Table 1.4 (*continued*)

No.	Structure	K_i (nM)[a]	
		plm II	cat D
17[d]		7.0	530
21[d]		62	56

[a]Average of >2 determinations, std error, <15%.
[b]Duplicate decode.
[c]Triplicate decode.
[d]Not a screening decode.
[e]Mixture of diastereomers, see Figure 2.

screening data. Inhibitors **9** and **10** possess K_is = 29 and 16 nM, respectively, against plm II and show modestly selective (1.5–2 ×) *vs* cat D: **9**: K_i = 44 nM, **10**: K_i = 32 nM. The selected resynthesized compounds for cat D also confirmed the activity and selectivity as observed in the primary screen. Inhibitor **11**, containing the D-Tiq residue, possesses a K_i = 3 nM, while compounds **12** and **13** possess K_is = 4 and 18 nM, respectively. Statines **11–13** are *ca.* 5–70-fold selective *vs* plm II, and are structurally unique among previously reported cathepsin D inhibitors.

6 CORROBORATING SCREENING STRUCTURE ACTIVITY RELATIONSHIP (SAR) DATA WITH LIBRARY QA ANALYSIS

When an encoded library is screened and found biologically active, a large number of beads are decoded revealing the structures of potentially active compounds. Between the 1902 decoded beads in the three QA runs and the 131 decodes obtained from screening, there were 16 overlapping structures identified. In each case, there was a 100% correlation between these biologically active compounds and a found (F) designation in the QA analysis adding an additional level of validation for the library QA analysis. Confirmation of biological activity comes from compound resynthesis and evaluation, and typically, only 5–10 compounds are resynthesized for this purpose. In general, compounds selected for resynthesis are those found multiple times (replicate structures) and are the most potent as per the primary screen (it is this strategy that was followed in the screening of library **1** described above). Upon confirmation of biological activity, a decision may be made to resynthesize additional compounds from the library, synthesize individual analogs, or prepare a follow-up optimization library. In any event, the decision regarding the design and synthesis of new compounds, will, in part, rely on the nascent SAR obtained from the original screen. It is, therefore important that such SAR information be reliable and of superior quality. This is the benefit of library QA analysis. By rendering information on the overall success of library synthesis and the performance of individual synthons, library QA can be utilized to corroborate and elevate the level of confidence in the screening SAR data.

For example, SAR information obtained from decoded structures indicates that plm II and cat D have strong preferences for hydrophobic **A** synthons. Comparison of the preferred **A** synthons against the performance of the same synthons in library QA (Figs. 1.5 and 1.7) reveals that all were strong performers. However, one synthon conspicuously absent from the screening data is the methylthioethylamine synthon **A6**. Given the fact that this synthon is neutral and hydrophobic, it might be expected to be among the decoded structures; however, this was not the case as discussed earlier. The QA data for **A6** showed a rather dubious performance (average p averaged for the three QA runs is 48.4%). This uncertainty suggests, at the very least, that compounds containing this synthon are likely to be under-represented in the follow-up single bead screen. In light of the QA data, statine **14** was synthesized. Statine **14**, a putative library member, is an analog of **9**, in which the o-methoxybenzylamine (**A14**) is replaced with methylthioethylamine (**A6**). Statine **14** possesses a $K_i = 12$ nM against plasmepsin II and a $K_i = 44$ nM against cathepsin D, and is thus similar in potency and selectivity to inhibitor **9** (Table 1.4).

Similarly, furfurylamine **A11** was not found in decoded structures, although the structurally analogous R- and S-tetrahydrofuranylmethylamines (**A8** and **A9**) were both present. Library QA analysis clearly shows uncertainty regarding the presence of this synthon in cleaved library compounds. As was the case for the synthon **A6**, putative library members containing **A11** in combination with preferred R_B amino acids, would be expected to be highly active. To test this hypothesis, statine **15** was prepared. This analog of **9** (**A14** in **9** exchanged for **A11** in **15**) was indeed found to be a potent aspartyl protease inhibitor: **15**: $K_i = 26$ nM, plm II; 46 nM, cat D. In the absence of QA data, it may have been erroneously concluded that synthons **A6** and **A11** were "inferior" **A** synthons.

As a final comparative example for the **A** synthons, positively charged, hydrophilic amines such as the pyridinylmethyl amines (**A17**–**A19**, **A36** and **A37**) and imidazolylpropylamine **A32** were not found in any of the decodes from the screen.[101,102] This suggests that compounds containing these synthons would be much weaker inhibitors ($>30\%$ control remaining) relative to those inhibitors containing the neutral, hydrophobic **A** amines. Inspection of the QA data for these synthons reveals uncertainty in their performance. In particular, **A19** ($p = 55.6\%$; QA run-3) and **A32** ($p = 33.3\%$; QA run-3) may be considered poor to mediocre synthons. Although the screening SAR strongly argues that compounds containing these synthons are unlikely to be as potent as their hydrophobic congeners, this information cannot be reliably established based on library screening. To obtain a more accurate portrait of the SAR, putative library members, statines **16** and **17** containing the 4-pyridinylmethyl (**A19**) and imdazoylylpropylamine (**A32**) synthons were synthesized. Remarkably, compounds **16** ($K_i = 15$ nM, plm II;

$K_i = 140$ nM, cat D) and **17** ($K_i = 7$ nM, plm II; $K_i = 530$ nM, cat D) were found to be the most potent and selective inhibitors of plm II identified in the library (Table 1.4). The ability of plm II to tolerate protonated A synthons *vs* cat D, leading to enhanced selectivity, is a salient SAR feature that would have been completely lost if we had relied solely on screening decode data.

One novel and unexpected discovery observed in screening, and verified by resynthesis, was the identification of D-Tiq (**B58**) as an amino acid residue selective for cathepsin D. This is exemplified by inhibitor **11**: $K_i = 3$ nM, cat D; $K_i = 210$ nM, plm II. Although all four isomeric tetrahydroquinoline carboxylic acids **B58–B61** were included in the library, only **B58** appeared in the decodes. Because it is uncommon to find D-amino acid residues located at the $P_{2'}$ position in protease substrates and inhibitors, the reliability of the screening SAR may be suspect. Examination of the library QA data for synthons **B58–B61** reveals that they were a successful group of synthons ($p > 85\%$). In this instance, the QA data reinforces the confidence in the SAR screening data in that D-Tiq is in fact the preferred synthon. This was substantiated with the synthesis and evaluation of inhibitors **18–20** (Table 1.4), the diastereomers of **11** where the **B58** D-Tiq residue was replaced with L-Tiq (**B59**), L- and D-Tic (**B50** and **B61**). Inhibitor **18** containing L-Tiq has a $K_i = 280$ nM against cat D, the L- and D-Tic containing inhibitors **19** and **20** possess inhibition constants of 290 and 2900 nM, respectively, against cat D. These data are clearly consistent with the activity observed in primary screening and corroborated by library QA.

Lastly, library QA indicates a virtual failure of synthon **B46**. In the absence of this data, it would have been concluded from screening that inhibitors containing **B46** would have potencies estimated to be > 25 nM, based on the $< 30\%$ of control cut-off at the 25 nM screening concentration. In reality, library members containing this synthon were probably never present in the screen. Hence, no conclusion may be drawn regarding the activity of library compounds containing the **B46** residue. Individual compounds containing this synthon must be synthesized separately in order to determine their biological profile. One compound (**21**) of this class was prepared, an analog of inhibitor **12**, substituting synthon **B46** for synthon **B45** (Table 1.4). Statine **12**, identified as a duplicate from the screen, is a potent, selective inhibitor of cat D (4 nM, 7-fold selective); compound **21** containing **B46** was found weakly active and non-selective: K_i for **21** equals 56 nM, cat D; K_i62 nM, plm II. Several other synthons including L- and D-asparagine (B19 and B20), L- and D-glutamine (B28 and B29), and L- and D-methionine (B32 and B33), and would likewise have to be synthesized as discrete compounds to more accurately establish the enzymes' affinity toward these amino acids.

7   NEW DEVELOPMENTS

The quality control procedures outlined in the preceding sections have resulted in an average single bead QA found rate of 80% for libraries produced at Pharmacopeia since late 1999. As with library **1**, the typical result is that "not found" cases are distributed rather evenly amongst the individual synthons at each step with occasional "problem synthons" displaying an uncharacteristically high failure rate. The problem synthons in library **1** (*i.e.* **46b**, Fig. 1.6) were not evident on simple analysis at an earlier stage of the library, but only when all chemistry had been completed. Thus, to detect the trouble, before the problem synthons were mixed with the good performers would be difficult. Samples of resin at each stage would need to be carried through a mock up of the remaining steps of the library synthesis and then analyzed. If the chosen synthons for the remaining steps happened to be ones that in combination elicit the failure in the synthon set being tested, the problem synthon(s) could be identified and replaced.

In other cases, however, individual synthon failures *are* evident by simple analysis at the stage the synthon is introduced. When this is the case, a simple "real time" in-line QA analysis of the resin sets at each stage of library production will identify problem synthons before they are thoroughly intermixed with the others. Historically, the predominant issue with introducing an in-line QA protocol has been the time needed to cleave and analyze both the tags and the compounds from a statistically relevant number of beads at each step. These delays affected work flow and library production efficiency and as a result, in-line library QA was not widely implemented. Recently, an automated method to perform the oxidative tag cleavage process in the same 96 well filter plate that the compound elution is performed has been developed. This advancement has shortened the time required to decode large numbers of beads by a factor of 5 and reduced the overall time needed for in-line library QA to approximately 4 days per analysis. Additionally, the custom data analysis program has been further refined to distinguish between "found (F; gray)", "questionable (Q; white)" and "not found (NF; black)." For both found and questionable, the XIC chromatogram reveals the expected mass $(M + H)^+$ and the signal to noise ratio is greater than three. In the case of the found answer, the peak height is within the linear range of the standard curve, while for the questionable result, the peak height is at or below the limit of quantitation. With these

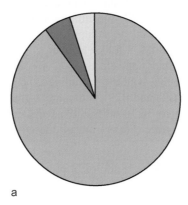

a

Figure 1.8   (a) Composite QA for library **2**-intermediate; (b) **A** synthon analysis for library **2**-intermediate; (c) **B** synthon analysis for library **2**-inermediate; (d) **C** synthon analysis for library **2**-intermediate.

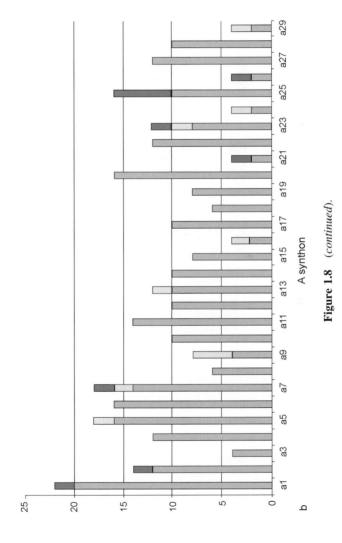

Figure 1.8 (*continued*).

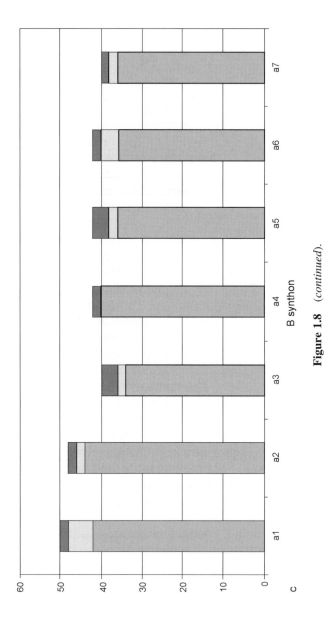

Figure 1.8 (*continued*).

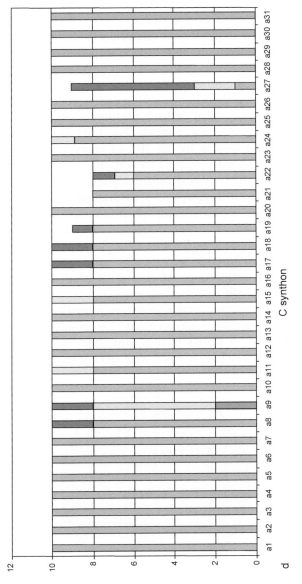

Figure 1.8 (*continued*).

improvements, in-line QA is now an integral part of the library quality control measures routinely employed at Pharmacopeia for library creation and compound elution, giving valuable chemical information.

Library 2 is a Pharmacopeia ECLiPS library prepared through four distinct chemical steps and in which the compounds are attached to the beads through an acid cleavable linker. During the synthesis of this library, the benefits of performing real time in-line QA were realized. After completing the first two library steps with uneventful in-line QA analysis, the library was again pooled and split and subjected to a series of chemical reactions comprising step 3. Following this, a total of 10 beads from each of the 31 distinct reaction vessels were selected for analysis. The analysis of 304 beads from this set, 6 beads lost during processing, resulted in 274 beads, or 90%, that were found to have produced compounds exhibiting the expected molecular weight for the synthesis (Fig. 1.8a). Figure 1.8b–d show the breakdown by step and synthon for the 10% of beads, which did not produce compounds of the expected molecular weight. R1 and R2 synthon analysis (Fig. 1.8b,c) show the fairly typical pattern of small percentages of not found compounds evenly distributed amongst all of the synthon choices. This indicates no gross synthon failure and corroborates the earlier QA results that steps 1 and 2 had no problems of this nature. R3 synthon analysis, however, clearly shows gross synthon failure for synthons 9 and 27 (Fig. 1.8d). Subsequent investigation of the synthon set revealed that these two were the only members of the group to contain highly electron-rich aromatic rings which are apparently unstable toward the strongly acidic conditions of compound cleavage.

Since progress on the synthesis of this library was paused for the results of this in-line QA, the vessels containing synthons 9 and 27 were simply removed from the library. The remaining 29 vessels were pooled, split and subjected to step 4 chemistry. Once again, a number of beads were taken from each of the 16 distinct

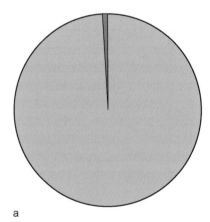

a

Figure 1.9   (a) Composite QA for library 2-final; (b) **A** synthon analysis for library 2-final; (c) **B** synthon analysis for library 2-final; (d) **C** synthon analysis for library 2-final; (e) **D** synthon analysis for library 2-final.

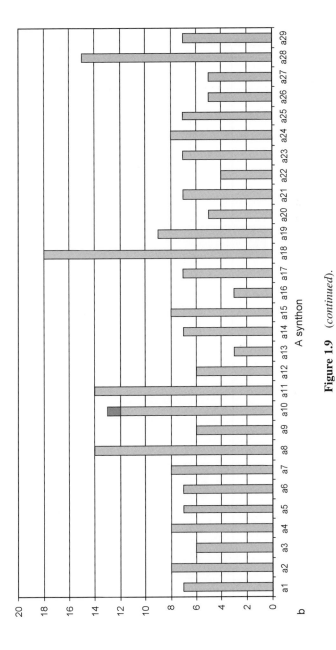

Figure 1.9 *(continued)*.

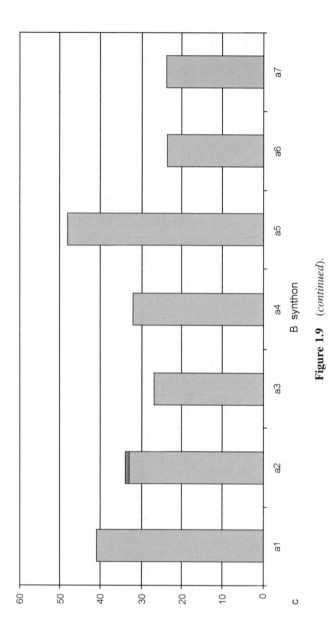

Figure 1.9 (*continued*).

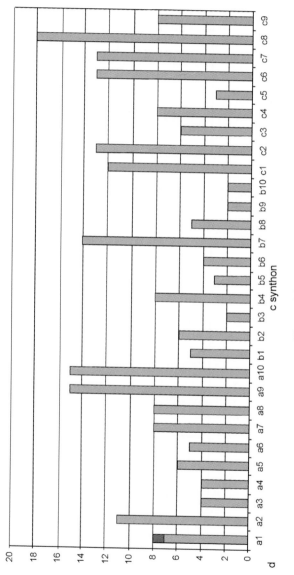

Figure 1.9 *(continued).*

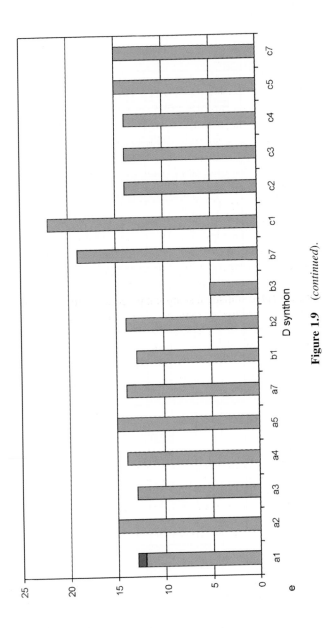

Figure 1.9 (*continued*).

step 4 reaction vessels and analyzed. These results (Fig. 1.9b–e) show that there were no problem synthons in step 4, and that with the removal of the two problem synthons from step 3, the overall library quality was dramatically improved to 99% found, with only one failed synthon from the expected synthetic steps (Fig. 1.9a).

8 CONCLUSION

Pharmacopeia has developed and put into routine practice a series of quality control protocols to ensure the highest degree of library fidelity. Quality control steps occur throughout the entire library construction pipeline. Extensive synthon profiling, rigorous analysis of QC compounds, final confirmation of library chemistry (library QA), and production-in-process-QC analysis are essential to ensure a successful library synthesis and production elution. Both in-line and final library QA provide empirical information regarding overall fidelity and an indication of the performance of individual synthons.

Library QA is a powerful analytical protocol to assess the quality of large, encoded combinatorial libraries. Based on the combined application of tag decoding and single bead LC/MS, QA analysis would be virtually impossible to carry out without an encoding strategy due to redundant masses produced during split synthesis. The statistical theory and its application to library assessment are analogous to the accepted statistical sampling practices used in industry to ascertain the quality of mass-produced material, e.g. tablet quality control in the pharmaceutical industry. The simplicity of tag reading and rapid acquisition of SAR information is arguably the most significant advantage of encoding technology vs other deconvolution techniques.[4–6] Library QA serves to substantiate and enhance the value of nascent SAR obtainment from library screening.

REFERENCES

1. R.E. Dolle, J. Comb. Chem., 2, 383–433 (2000).
2. R.E. Dolle and K.H. Nelson, Jr., J. Comb. Chem., 1, 235–282 (1999).
3. R.E. Dolle, Mol. Divers., 3, 199–233 (1998).
4. S.E. Schneider and E.V. Anslyn, Adv. Supramol. Chem., 5, 55–120 (1999).
5. A. Furka and W.D. Bennett, Comb. Chem. High Throughput Screen, 2, 105–122 (1999).
6. J.J. Baldwin and R.E. Dolle, Annu. Rep. Comb. Chem. Mol. Divers., 1, 287–297 (1997).
7. J.J. Baldwin, Mol. Divers., 2, 81–88 (1996).
8. J.C. Chabala, Curr. Opin. Biotechnol., 6, 632–639 (1995).
9. M.H.J. Ohlmeyer, R.N. Swanson, L.W. Dillard, J.C. Reader, G. Asouline, R. Kobayashi, M. Wigler, and W.C. Still, Proc. Natl Acad. Sci. USA, 90, 10922 (1993).
10. A. Borchardt and W.C. Still, J. Am. Chem. Soc., 116, 373–374 (1994).
11. P. Seneci, S. Miertus, and G. Fassina (eds), Combinatorial Chemistry and Technology: Principles, Methods, and Applications, Marcel Dekker, Inc., New York, pp. 127–167 (1999).
12. M.S. Shchepinov, R. Chalk, and E.M. Southern, Tetrahedron, 56, 2713–2724 (2000).

13. R.H. Scott, C. Barnes, U. Gerhard, and S. Balasubramanian, *Chem. Commun.*, **14**, 1331–1332 (1999).
14. C. Barnes, R.H. Scott, and S. Balasubramanian, *Recent Res. Dev. Org. Chem.*, **2**, 367–379 (1998).
15. J.E. Hochlowski, D.N. Whittern, and T.J. Sowin, *J. Comb. Chem.*, **1**, 291–293 (1999).
16. I. Hughes, *J. Med. Chem.*, **41**, 3804–3811 (1998).
17. W.L. Fitch, T.A. Baer, W. Chen, F. Holden, C.P. Holmes, D. Maclean, N. Shah, E. Sullivan, M. Tang, P. Waybourn, S.M. Fischer, C.A. Miller, and L.R. Snyder, *J. Comb. Chem.*, **1**, 188–194 (1999).
18. D.S. Wagner, C.J. Markworth, C.D. Wagner, F.J. Schoenen, C.E. Rewarts, B.K. Kay, and H.M. Geysen, *Comb. Chem. High Throughput Screening*, **1**, 143–153 (1998).
19. X.-U. Xiao and M.P. Nova, *Comb. Chem.*, 135–152 (1997).
20. S.S. Rahman, D.J. Busby, and D.C. Lee, *J. Org. Chem.*, **63**, 6196–6199 (1998).
21. A.W. Czarnik, *Curr. Opin. Chem. Biol.*, **1**, 60–66 (1997).
22. M.H. Geysen, C.D. Wagner, W.M. Bodnar, C.J. Markworth, G.J. Parke, F.J. Schoenen, D.S. Wagner, and D.S. Kinder, *Chem. Biol.*, **3**, 679–688 (1996).
23. J.W. Guiles, C.L. Lanter, and R.A. Rivero, *Angew. Chem. Int. Ed.*, **37**, 926–928 (1998).
24. B.J. Battersby, D. Bryant, W. Meutermans, D. Matthews, M.L. Smythe, and M. Trau, *J. Am. Chem. Soc.*, **122**, 2138–2139 (2000).
25. X. Xiao, C. Zhao, H. Potash, and M.P. Nova, *Angew. Chem. Int. Ed. Engl.*, **36**, 780–782 (1997).
26. S. Lane and A. Pipe, *Rapid Commun. Mass Spectrom.*, **13**, 798–814 (1999).
27. Z.J. Ni, D. Maclean, C.P. Holmes, M.M. Murphy, B. Ruhland, J.W. Jacobs, E.M. Gordon, and M.A. Gallop, *J. Med. Chem.*, **39**, 1601–1608 (1996).
28. D. Maclean, J.R. Schullek, M.M. Murphy, Z.J. Ni, E.M. Gordon, and M.A. Gallop, *Proc. Natl Acad. USA*, **94**, 2805–2810 (1997).
29. J.L. Silen, A.T. Lu, D.W. Solas, M.A. Gore, D. Maclean, N.H. Shah, J.M. Coffin, N.S. Bhinderwala, Y. Wang, K.T. Tsutsui, G.C. Look, D.A. Campbell, R.L. Hale, M. Navre, and C.R. DeLuca-Flaherty, *Antimicrob. Agents Chemother.*, **42**, 1447–1453 (1998).
30. H.M. Geysen, C.D. Wagner, W.M. Bodnar, C.J. Markworth, G.J. Parke, F.J. Schoenen, D.S. Wagner, and D.S. Kinder, *Chem. Biol.*, **3**, 679–688 (1996).
31. P.A. Keifer, *Drug Discov. Today*, **2**, 468–478 (1997).
32. A.W. Schwabcher, Y. Shen, and C.C. Johnson, *J. Am. Chem. Soc.*, **121**, 8669–8670 (1999).
33. J.M. Kerr, S.C. Banville, and R.N. Zuckerman, *J. Am. Chem. Soc.*, **115**, 2529–2531 (1993).
34. S. Brenner and R.A. Lerner, *Proc. Natl Acad. USA*, **89**, 5381–5383 (1992).
35. A. Furka, J.W. Christensen, E. Healy, H.R. Tanner, and H. Saneii, *J. Comb. Chem.*, **2**, 220–223 (2000).
36. J.J. Baldwin, J.J. Burbaum, I. Henderson, and M.H.J. Ohlmeyer, *J. Am. Chem. Soc.*, **117**, 5588–5589 (1995).
37. J.C. Chabala, J.J. Baldwin, J.J. Burbaum, D. Chelsky, L. Dillard, I. Henderson, G. Li, M.H.J. Ohlmeyer, and T.L. Randle, *Perspect. Drug Discov. Des.*, **2**, 305–318 (1995).
38. J.J. Baldwin, J.J. Burbaum, D. Chelsky, L.W. Dillard, I. Henderson, G. Li, M.H.J. Ohlmeyer, T.L. Randle, and J.C. Reader, *Eur. J. Med. Chem.*, **30**, 349s–358s (1995).
39. J.J. Baldwin and I. Henderson, *Med. Res. Rev.*, **16**, 391–405 (1996).
40. J.J. Baldwin, *Mol. Divers.*, **2**, 81–88 (1996).
41. J.J. Burbaum, M.H.J. Ohlmeyer, J.C. Reader, I. Henderson, L.W. Dillard, G. Li, T.L. Randle, N.H. Sigal, D. Chelsky, and J.J. Baldwin, *Proc. Natl Acad. Sci. USA*, 6027–6031 (1995).
42. L. Rokosz, N.H. Sigal, J.J. Burbaum, D. Chelsky, L.W. Dillard, I. Henderson, G. Li, M.H.J. Ohlmeyer, T.L. Randle, and J.C. Reader, *Innovation Perspect. Solid Phase Synth. Comb. Libr., Collect. Pap. Int. Symp. 4th Meeting Date 1995*, Mayflower Scientific, Birmingham, pp. 101–106 (1996).

43. J.J. Baldwin and I. Henderson, In: *High Throughput Screening: The Discovery of Bioactive Substances*, (J.P. Devlin, ed.), Marcel Dekker, Inc., New York, pp. 167–190 (1997).
44. J.J. Baldwin, In: *Combinatorial Chemistry and Molecular Diversity in Drug Discovery*, (E.M. Gordon and J.F. Kerwin, Jr. eds), Wiley-Liss, Inc., New York, pp. 181–188 (1998).
45. K. McMillan, M. Adler, D.S. Auld, J.J. Baldwin, E. Blasko, L.J. Browne, D. Chelsky, D. Davey, R.E. Dolle, K.A. Eagen, S. Erickson, R.I. Fledman, C.B. Glaser, C. Mallari, M.M. Morrissey, M.H.J. Ohlmeyer, G. Pan, J.F. Parkinson, G.B. Phillips, M.A. Polokoff, N.H. Sigal, R. Vergona, M. Whitlow, T.A. Young, and J.J. Devlin, *Proc. Natl Acad. USA*, **97**, 1506–1529 (2000).
46. R.A. Horlick, M.H. Ohlmeyer, I.L. Stroke, B. Strohl, G. Pan, A. Schilling, V. Paradkar, J.G. Quintero, M. You, C. Riviello, M.B. Thorn, B. Damaj, V.D. Fitzpartick, R.E. Dolle, M.L. Webb, J.J. Baldwin, and N.H. Sigal, *Immunopharmacology*, **43**, 169–177 (1999).
47. J.J. Baldwin, G.L. Kirk, L.W. Dillard, and J.A. Connelly, *Methods Mol. Cell. Biol.*, **6**, 74–88 (1996).
48. J.J. Baldwin and E.G. Horlbeck, US Patent 5,663,046 (1997).
49. C.A. Srebalus and J. Li, *Am. Soc. Mass Spectrom.*, **11**, 352–355 (2000).
50. J.A. Loo, *Eur. J. Mass Spectrom.*, **3**, 93–104 (1997).
51. J.P. Nawrocki, M. Wigger, C.H. Watson, T.W. Hays, M.W. Senko, S.A. Benner, and J.R. Eyler, *Rapid Commun. Mass Spectrom.*, **10**, 1860–1864 (1996).
52. Y.M. Dunyevskiy, P. Vouros, E.A. Wintner, G.W. Shipps, T. Carell, and J. Rebek, Jr., *Proc. Natl Acad. Sci. USA*, **93**, 6152–6157 (1996).
53. S.C. Promerantz, J.A. McCloskey, T.M. Tarasow, and B.E. Eaton, *J. Am. Chem. Soc.*, **19**, 3861–3867 (1997).
54. J.W. Metzger, C. Kempter, K.-H. Wiesmuller, and G. Jung, *Anal. Biochem.*, **219**, 261–277 (1994).
55. C.L. Brummel, I.N.W. Lee, Y. Zhou, S.J. Benkovic, and N. Winograd, *Science*, **264**, 399–401 (1994).
56. K. Berlin, R.J. Jain, C. Tetzlaff, C. Steinback, and C. Richert, *Chem. Biol.*, **4**, 63–77 (1997).
57. R.E. Dolle, J. Guo, W. Li, N. Zhao, and J.A. Connelly, *Mol. Divers.*, **5**(1), 35–49 (2001).
58. R.E. Dolle, J. Guo, L. O'Brien, Y. Jin, M. Piznik, K.J. Bowman, W. Li, W.J. Egan, C.L. Cavallaro, A.L. Roughton, W. Zhao, J.C. Reader, M. Orlowski, B. Jacob-Samuel, and C. Dilanni-Carroll, *J. Comb. Chem.*, **2**(6) (2000).
59. R.E. Dolle, In: *Optimization of Solid-Phase Combinatorial Synthesis*, (B. Yan and A.W. Czarnik eds), Marcel Dekker, Inc., New York (2001).
60. K.C. Lewis, W.L. Fitch, and D. Maclean, *LC–GC*, **16**, 644–649 (1998).
61. B.J. Egner and M. Bradley, *Drug Discov. Today*, **2**, 102–109 (1997).
62. M.A. Gallop and W.L. Fitch, *Curr. Opin. Biol.*, **1**, 94–100 (1997).
63. W.L. Fitch, *Mol. Divers.*, **4**, 39–45 (1999).
64. L.R. Snyder, J.J. Kirkland, and J.L. Glajch, *Practical HPLC Method Development*, 2nd ed., Wiley, New York (1997).
65. J.N. Kyranos and J.C. Hogan, Jr., *Mod. Drug Discov.*, **2**, 73–81 (1999).
66. J.N. Kyranos and J.C. Hogan, Jr., *Anal. Chem.*, **70**, 389A–395A (1998).
67. M. Patek, B. Drake, and M. Labl, *Tetrahedron Lett.*, **36**, 2227–2230 (1995).
68. J.M. Ostresh, G.M. Husar, S.E. Blondelle, B. Dorner, P.A. Weber, and R.A. Houghten, *Proc. Natl Acad. Sci. USA*, **91**, 11138–11142 (1994).
69. S.M. Hutchins and K.T. Chapman, *Tetrahedron Lett.*, **36**, 2583–2586 (1995).
70. A.M. Bray, D.S. Chiefari, R.M. Valerio, and N.J. Maeji, *Tetrahedron Lett.*, **36**, 5081–5084 (1995).
71. C.E. Kibbey, *Mol. Divers.*, **1**, 247–258 (1995).

72. B.H. Hsu, E. Orton, S.-Y. Tang, and R.A. Carlton, *J. Chromatogr. B*, **725**, 103–112 (1999).
73. P.A. Asmus and J.B. Landis, *J. Chromatogr.*, **316**, 461–472 (1984).
74. M. Lebl, V. Krchnak, G. Ibrahim, J. Pires, C. Burger, Y. Ni, Y. Chen, D. Podue, P. Mudra, V. Pokorny, P. Poncar, and K. Zenisek, *Synthesis*, **11**, 1971–1978 (1999).
75. L. Fang, M. Wan, M. Pennacchio, and J. Pan, *J. Comb. Chem.*, **2**, 254–257 (2000).
76. J.-F.A. Borney and M.E. Homen, *LC–GC*, **18**, S14–S19 (2000).
77. E.W. Taylor, M.G. Qian, and G.D. Dollinger, *Anal. Chem.*, **70**, 3339–3347 (1998).
78. E.M. Fujinari and L.O. Courthaudon, *J. Chromatogr. A*, **592**, 209–214 (1992).
79. A.B. Sage, D. Little, and K. Giles, *LC–GC*, **18**, S20–S29 (2000).
80. R.D. Sussmuth and G. Jung, *J. Chromatogr. B*, **725**, 49–65 (1999).
81. P.S. Marshall, *Rapid Commun. Mass Spectrom.*, **13**, 778–781 (1999).
82. B.D. Dulery, J. Verne-Mismer, E. Wolf, C. Kugel, and L. Van Hijfte, *J. Chromatogr. B*, **725**, 39–47 (1999).
83. M.C. Fitzgerald, K. Harris, C.G. Shevlin, and G. Siuzdak, *Bioorg. Med. Chem. Lett.*, **6**, 979–982 (1996).
84. R. Richmond, E. Gorlach, and J.-M. Seifert, *J. Chromatogr. A*, **835**, 29–39 (1999).
85. R. Richmond and E. Gorlach, *Anal. Chim. Acta*, **390**, 175–183 (1999).
86. R. Richmond and E. Gorlach, *Anal. Chim. Acta*, **394**, 33–42 (1999).
87. G. Hegy, E. Gorlach, R. Richmond, and R. Bitsch, *Rapid Commun. Mass Spectrom.*, **10**, 1894–1900 (1996).
88. P.-H. Lambert, J.A. Boutin, S. Bertin, J.-L. Fauchere, and J.-P. Volland, *Rapid Commun. Mass Spectrom.*, **11**, 1971–1976 (1997).
89. J.W. Metzger, J.W. Kempler, K.-H. Wiesmiller, and G. Jung, *Anal. Biochem.*, **219**, 261–277 (1994).
90. P.A. Keifer, S.H. Smallcombe, E.H. Williams, K.E. Salomon, G. Mendez, J.L. Belletire, and C.D. Moore, *J. Comb. Chem.*, **2**, 151–171 (2000).
91. L. Schultz, C.D. Garr, L.M. Cameron, and J. Bukowski, *Bioorg. Med. Chem. Lett.*, **8**, 2409–2414 (1998).
92. L. Zeng, L. Burton, K. Yung, B. Shushan, and D.B. Kassel, *J. Chromatogr. A*, **794**, 3–13 (1998).
93. S.K. Thompson, *Sampling*, Wiley, New York (1992).
94. D.A. Berry and B.W. Lindgren, *Statistics: Theory and Methods*, Brooks/Cole Publishing Company, Pacific Grove (1990), Chapter 4.
95. C.J. Clopper and E.S. Pearson, *Biometrika*, **26**, 404–413 (1934).
96. A.M. Silva, A.Y. Lee, S.V. Gulnik, P. Majer, J. Collins, T.N. Bhat, P.J. Collins, R.E. Cachau, K.E. Luker, I.Y. Gluzman, S.E. Francis, A. Oksman, D.E. Goldberg, and J.W. Erickson, *Proc. Natl Acad. Sci. USA*, **93**, 10034–10039 (1996).
97. R.P. Moon, L. Tyas, U. Certa, K. Rupp, D. Bur, C. Jacquet, H. Matile, H. Loetscher, F. Grueninger-Leitch, J. Kay, B.D. Dunn, C. Berry, and R.G. Ridley, *Eur. J. Biochem.*, **244**, 552–560 (1997).
98. C.D. Carroll, H. Patel, T.O. Johnson, T. Guo, M. Orlowski, Z.-M. He, C.L. Cavallaro, J. Guo, A. Oksman, I.Y. Gluzman, J. Connelly, D. Chelsky, D.E. Goldbert, and R.E. Dolle, *Bioorg. Med. Chem. Lett.*, **8**, 2315–2320 (1998).
99. C.D. Carroll, T.O. Johnson, S. Tao, G. Lauri, M. Orlowski, I.Y. Gluzman, D.E. Goldbert, and R.E. Dolle, *Bioorg. Med. Chem. Lett.*, **8**, 3203–3206 (1998).
100. T.S. Haque, G. Skillman, C.E. Lee, H. Habashita, I.Y. Gluzman, T.J.A. Ewing, D.E. Goldberg, I.D. Kuntz, and J.A. Ellman, *J. Med. Chem.*, **42**, 1428–1440 (1999).
101. R.E. Dolle, *Abstracts of Papers*, 218th National Meeting of the American Chemical Society, New Orleans, LA, American Chemical Society, Washington, DC (1999), MEDI 13.
102. J. Guo, C. Hicks, N. Zhao, W.J. Egan, G. Lauri, C.L. Cavallaro, and R.E. Dolle, *Abstracts of Papers*, 218th National Meeting of the American Chemical Society, New Orleans, LA, American Chemical Society, Washington, DC (1999), ORGN.

2

Analysis of a Combinatorial Library Synthesized Using a Split-and-Pool Irori MicroKan Method for Development and Production

Heewon Lee

Chemical Development Department, Boehringer Ingelheim Pharmaceuticals, Inc., 900 Ridgebury Road, Ridgefield, CT 06877, USA

Christopher R. Sarko

Medicinal Chemistry Department, Boehringer Ingelheim Pharmaceuticals, Inc., 900 Ridgebury Road, Ridgefield, CT 06877, USA

CONTENTS

High Throughput Analysis for Early Drug Discovery
Edited by James N. Kyranos
© 2004 Elsevier Inc. All rights reserved

1 INTRODUCTION

The advent of split-and-pool synthesis has had a dramatic effect upon the ability of the chemist to produce large combinatorial libraries. These multi-dimensional libraries can be quickly assembled from small sets of reagents producing very large libraries.[1-3] The use of chemical tagging techniques in split-and-pool synthesis was a method devised to prevent the time-consuming deconvolution steps required in early syntheses.[4] Furthermore, chemical tags may not be compatible with all relevant chemistries one could perform on a routine basis. Alternatives to chemical tags reported in the literature include color or shape encoding strategies.[5-8] Another choice is radio frequency (RF) tagging microchips. These systems implore the use of a binary encoded transponder contained within glass.[9,10] One such commercially marketed system is the Irori AccuTag system that has been implemented at Boehringer Ingelheim Pharmaceuticals Inc.

Another common issue with split-and-pool synthesis is the formation of complex mixtures, once again requiring deconvolution. The use of a discrete container for solid supports is one potential method to prevent the mixture deconvolution, however, identification of these containers would be difficult. Incorporation of a discrete tag such as in the AccuTag system allows for individual compound generation while tracking compound identity. Such a synthetic vessel is the Irori MicroKan; this vessel can hold 30 mg of solid support and encase the RF tag, thus combining identity and individuality of the compounds. Another advantage of this type of system is that one can use traditional glassware in a more parallel approach. Therefore, combinatorial libraries generated by the Irori MicroKan method benefit not only from the split-and-pool synthesis on solid support but also from the parallel synthesis technique producing one target compound per well without the need for deconvolution.

As library synthesis utilizes automated split-and-pool parallel synthesis technology, the number of compounds generated can easily outnumber the throughput that serial analytical techniques can provide unless multiple instruments are employed simultaneously. The analytical challenge is to confirm the synthesis of target products and, ideally, assess the purity for all the compounds produced. Moreover, the processing and the intuitive representation of the huge amount of analytical data could become problematic.

There are several reported high throughput analytical methods to accomplish combinatorial library characterization. These methods include flow-injection mass spectral analysis,[11-13] ultrafast LC methods,[14-16] parallel LC,[17,18] and supercritical fluid chromatography (SFC).[19] A few articles reviewing high throughput analysis techniques are also available in the literature.[20-22] As the number of compounds generated by parallel synthesis expands, rapid purification of the compounds becomes the bottleneck of the process. Traditional chromatographic purification or aqueous extraction workup for individual sample mixtures are usually not manageable for combinatorial libraries. To achieve compound purification in an automated and parallel fashion, researchers have employed several methods including resin scavengers,[23] solid phase extraction (SPE),[24] and solid-supported liquid–liquid extraction (SLE).[25,26]

In this chapter, we discuss the analytical methods and techniques used in the combinatorial chemistry group for 100% quality control (QC) analysis during the development and production of the libraries. Automated SPE methods as high throughput purification of combinatorial libraries are also described.

2 INSTRUMENTATION

All analyses were performed using high-performance liquid chromatography (HPLC) coupled with ultraviolet (UV) absorbance, evaporative light scattering detection (ELSD), and mass spectrometry (MS). Two systems were developed for the purpose of 100% QC of library compounds; a traditional single-channel HPLC/UV/ELSD/MS system, and a multiplexed HPLC/UV/ELSD/MS system that was capable of analyzing four samples simultaneously for high throughput. A fast generic LC gradient method was developed to analyze wide variety of compound classes and was used as a standard method for all compounds, unless validation plates during the library development required a different separation method.

2.1 Single-Channel HPLC/UV/ELSD/MS System

A liquid chromatography system based on an Agilent 1100 (Palo Alto, CA) quaternary pump with a degasser and an 1100 DAD UV detector was used as the platform. Gilson 215 liquid handler (Middleton, WI) was used as an autosampler with a single injection needle. A Sedex 55 ELSD was purchased from S.E.D.E.R.E (Alfortville, France). Mass spectrometric detection was performed by using a Micromass (Manchester, UK) Platform II single quadrupole mass spectrometer.

For the generic gradient method, a Zorbax SB-C8 Rapid Resolution cartridge column (Agilent Technologies, 4.6 mm i.d. × 30 mm length) was employed at a 2.5 ml/min flow rate. The HPLC grade solvents employed, water and acetonitrile, were purchased from EM Science (Gibbstown, NJ). To help ionization processes in the mass spectrometer the solvents were acidified with 0.1% v/v formic acid (EM Science).

The initial pump setting for the LC gradient was 85% of mobile phase A (water with 0.1% formic acid) and 15% of mobile phase B (acetonitrile with 0.1% formic acid) at a flow rate of 2.5 ml/min. A linear gradient ramp was applied from 15 to 95% B in 1.75 min and then the pump was held at 95% B until 2.0 min. By 2.1 min the pump was ramped down to its initial condition until the next gradient. Minimization of tubing length and tubing internal diameter was critical to decrease system dead volume for fast gradient methods. By using a short column the total run time was 3.2 min per sample from injection to injection with a throughput of 450 samples/day (over four 96-well plates/day). The flow was split ca. 10:1 between ELSD and MS subsequent to UV detection. The ELSD was operated at 40°C and at a 4.0 bar of nitrogen nebulization flow. The mass spectrometer was operated in either electrospray (ES) ionization or atmospheric pressure chemical ionization (APCI) mode depending on sample requirements. UV absorbance was measured at either 220 or 254 nm depending on library chromophore characteristics.

2.2   Four-Channel Multiplexed HPLC/UV/ELSD/MS System

This system was designed to analyze four samples simultaneously and was equipped with four-syringe autosamplers, four columns, four UV detectors, one mass spectrometer with a four-channel multiplexed electrospray interface (MUX),[27] and four ELS detectors as shown in Fig. 2.1. A Gilson 215 liquid handler with four syringes and four injection needles was used for sample injection. The LC system consisted of an Agilent 1100 quaternary pump with a degasser and four 1100 MWD UV detectors. Four Zorbax SB-C8 Rapid Resolution cartridge columns were used. Four Sedex 75C ELSDs were purchased from S.E.D.E.R.E. and Micromass LCT equipped with four-channel MUX was used for mass spectral detection.

The gradient conditions employed were the same as for the single-channel system, except that the pump flow rate was 10 ml/min and the flow was split into four at the Gilson 215 injector using two stages of three stainless steel tees. Every effort was made to make the four flow paths exactly the same by using the identical model UV and ELS detectors, employing tubing with the same i.d. and length for connections, and utilizing the same model columns. All four columns were replaced at the same time for regular maintenance and if column-related problems occurred.

Figure 2.1   Picture of multiplexed HPLC/UV/ELSD/MS system.

The total run time from injection to injection was 3.5 min with the throughput of 1600 samples/day (over sixteen 96-well plates/day). With MUX option, the LCT mass spectrometer could only be operated in ES ionization mode (positive or negative). The parameters for UV and ELSD were the same as those of the single channel system.

The mass spectrometer was required to acquire data from eight analog channels (four UVs and four ELSDs), however, the number of usable analog channel input ports provided for four channel MUX option was four by default. A few modifications were necessary to acquire two analog inputs for each channel. The initialization files for the LCT instrument were re-coded in a way that they instructed the LCT instrument that there were eight analog channels available, which would normally be used with eight-channel MUX option. By doing this, the instrument assigned two analog data input ports per sample channel. Also, the number of analog data points per second had to be decreased from the default value of 20 to 8 in order to have the MassLynx acquire all related data properly.

3   DATA PROCESSING

All UV, ELSD, and MS data were acquired using MassLynx software (version 3.5, Micromass). After acquisition, the data were automatically processed using OpenLynx (Micromass). As the sample flow reached the UV cell first, followed by splitting between MS and ELSD, a compound would be detected by UV first and then detected by MS and ELSD. OpenLynx software first aligned UV and ELSD traces with MS data using user-defined offset, so that the chromatographic UV, ELSD, and MS peaks of the same origin would have the same apparent retention time. Then, OpenLynx integrated the chromatographic traces, and a mass spectrum was generated for each integrated peak. Within the mass spectrum the ion of interest was searched and the target ion intensity was compared with the user-defined criteria. If the target ion intensity was greater than the user-defined criteria, then it was considered that the target ion was present and the product was "found" for the sample. The "purity" of the compound was defined as the percent peak areas of the corresponding UV and ELSD peaks. In this case, the chromatographic peak was colored green to indicate that the target product was found, and the sample location in the plate view was colored green. Otherwise, the sample location in the plate view was colored red to show that the product was "not found." For purity assessment, the solvent front was excluded from the data processing.

4   AUTOMATED SPE SYSTEM IN 96-WELL PLATE FORMAT

An SPE 215 system (Gilson) with a 735 Sampler software was used for high throughput purification of the libraries. It was equipped with eight syringes and eight probes so that eight samples (one entire column of 96-well plate format) could be handled simultaneously. The switching valve could control up to eight different solvents. Up to three 96-well sample plates could be placed per batch, as the instrument bed accommodated three collection plates. Depending on the nature of products and side-products/impurities, various SPE sorbents in the Versaplate 96-well format were purchased from Varian (Harbor City, CA), including Hydromatrix for SLE.[25] A customized SPE rack was developed by Gilson to accommodate Varian Versaplate. As a rule of thumb, the SPE cartridge sorbent size should be 10 times of the crude sample amount. In our case, a typical SPE cartridge size used for our library purification was 100 mg.

5   RESULTS AND DISCUSSION

The goal of library assessment and characterization was two-fold; one was to confirm the presence of the target compound, and the other was to measure compound purity. Quantitation of each compound was accomplished by weighing the final product in microtube format using Bohdan Automation weighing station. Since purified compounds for calibration were not available for each member of the combinatorial libraries, whose entities were most likely to be synthesized for the first time, purity estimation had to be performed without established standards most of the cases. Therefore, chromatographic purity assessment required a "universal" detection technique that responded to all classes of compounds. Low-wavelength UV detection is considered to be a general detection technique.[12-15] However, the detector responds to compounds with chromophore and its sensitivity is dependent on the extinction coefficients. Thus, compounds with strong chromophores would be over-represented and compounds without significant chromophores would be under-represented.

Another detection technique, evaporative light scattering detection (ELSD), was employed to estimate the compound purity on-line with UV. ELSD is based on light scattering from solute particles and is reported to provide nearly equivalent responses to compounds of similar classes.[28] Quantitation based on ELSD showed smaller quantitation error than UV-based data through compound-independent calibration approach.[29] Since ELSD detects light scattered by solute particles, volatile and low molecular weight (MW) compounds that completely evaporate (i.e. no particles formed) under the experimental conditions will not be registered by ELSD. It also should be noted that ELSD is a non-linear detector,[30,31] in contrast to linear UV detection. If the mass of a compound being detected were plotted against the ELSD signal peak area, it would appear as an exponential curve as shown in Fig. 2.2a. In other words, if X amount of sample produced Y response peak area, $2X$ amount would produce greater than $2Y$ response when comparing peak area vs peak area directly. UV-based detection generates linear calibration curve, while calibration curve for ELSD is typically plotted in log–log scale in order to obtain straight line relationship (as in Fig. 2.2b). The consequence is that the direct comparison of the ELSD peak areas for purity estimation, without calibration standards and without a calibration curve, could be biased toward the component with the greater peak area. Both UV and ELSD data were used to estimate purity of

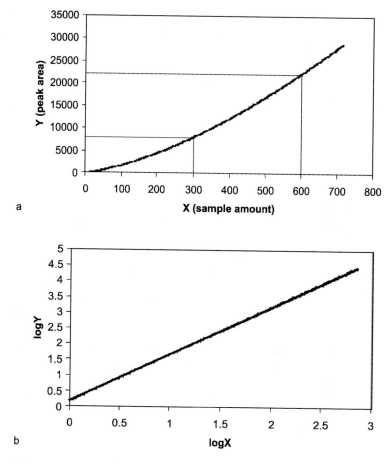

Figure 2.2   Non-linear characteristic of ELS detection: (a) direct plot of peak area *vs* sample amount with a correlation of $Y = AX^B$; (b) log–log plot of peak area *vs* sample amount with a correlation of $\log Y = \log A + B \log X$.

the libraries in this report. The two detection techniques, UV and ELSD, observe different properties of molecules, therefore, the "purity" information obtained could vary from each other.

To ensure analytical instrument performance, a three-component QC standard mixture was developed in order to verify chromatography and mass spectrometer performance. The compounds for the standard mixture were chosen so that the mixture provides varying degree of polarity thus displaying a range of LC retention times on our standard LC method. The compounds should ionize and be detected easily in the mass spectrometer using ES ionization and/or APCI mode with the intention that these chemicals can be used to test instrument performance. In addition, the MWs of the standards should be within the range of our actual samples (MW 200–700). The standard mixture was composed of reserpine (Sigma, St. Louis, MO), coumarin 314 (Aldrich, Milwaukee, WI), and 2-(2H-benzotriazol-2-yl)-4,6-di-*tert*-pentylphenol (Aldrich). Figure 2.3 shows the UV/ELSD/MS-TIC (positive ES ionization) traces of the QC standard acquired by one of the four-channels of the multiplexed system. The traces have been aligned so that each component would have the same apparent elution time for all three traces. The peak at 0.22 min was the solvent front. As dimethylsulfoxide (DMSO, used as the solvent for the QC standard mixture) absorbs UV at 220 nm, the UV trace showed a strong solvent front peak; while ELSD displayed no discernable peak for the solvent front as the solvent was evaporated forming no particles under the experimental conditions. The peaks with retention times of 0.94 and 1.21 min exhibited responses from all three detection modes, though the relative intensities were not the same. The third component, at retention time of 2.06 min on UV, did not ionize significantly by positive ES ionization. Mass spectral detection strongly depends on the functional groups in the compounds and on the ionization modes employed.[32] Because of this selectivity and specificity, MS is typically not suitable for general purity assessment. Figure 2.3 clearly illustrates the difficulties associated with purity assessment without a calibration standard, as can be seen in the different

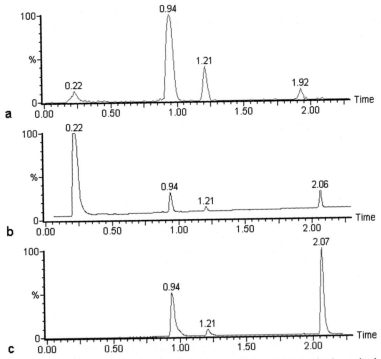

Figure 2.3 LC/UV/ELSD/MS data of three-component QC standard acquired by four-channel multiplexed system: (a) MS-total ion chromatogram (TIC); (b) UV trace at 220 nm; (c) ELSD trace.

relative peak intensities among the detection methods. Relying on single detection technique could be misleading. By employing two simultaneous detection methods for purity evaluation, UV and ELSD, possible bias toward or against a specific compound(s) could be minimized, though not entirely eliminated.

6 LIBRARY DEVELOPMENT STAGE

The essential elements of an AccuTag synthesis are reactors (MicroKans), RF tags, a RF transponder or base station, and a software package capable of tracking binary encoded RF tags. In Fig. 2.4, an RF tag, a MicroKan (with an RF tag inside), and a MiniKan are shown. MiniKans were used as tracing cans during our library production stage as explained in the following section. Once these key elements have been obtained the synthesis itself, while producing multi-dimensional libraries, was treated more as a parallel synthesis. We will use case examples of libraries detailing the steps and procedures used for reaction validation, library synthesis, and library cleavage.

The initial phase of library development was the creation of validation plate(s). These plates were generated to ensure that all reagents would react as required and that each reagent produced an acceptable yield of the desired product. These validation plates were not combinatorial mixtures; in order to test each combination of reagents one would have to produce the entire library, which would not be a very cost-effective method for validation. Rather, each reagent X was tested once against reagent Y, and this combination was tested against one element from reagent set Z. In this manner each reagent would be tested once for reaction completion and compatibility.

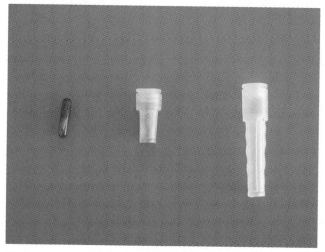

Figure 2.4 Picture of a radio frequency (RF) tag, an Irori MicroKan, and a tracing can.

These validation reactions were run under conditions similar to those expected in the final library, and each reaction was performed in the MicroKan to ensure that the Irori synthesis method was compatible with the desired chemical reactions. After cleavage of the material from the solid support, each compound was analyzed using LC/MS to determine which, if any, of the reagents needed to be removed or conditions modified. Only after completing this process was the library advanced to production status.

In our 3D library example with a biphenyl scaffold, the reagents need to be tested were 32 primary amines, 3 acids, and 44 primary/secondary amines. The total number of compounds for this library was 4224 ($32 \times 3 \times 44$). The validation consisted of four 96-well plates. The 32 primary amines were positioned from row A through H of the four plates (8×4). The 44 primary/secondary amines were arranged from column 1 through 11 of the four plates (11×4). Column 12 on each plate was reserved for analytical QC blanks and standards. The three acids were tested in plate 1, 2, and 3, respectively and plate 4 was synthesized with acid 1 again. The analytical data for these four validation plates were presented in Fig. 2.5, generated using OpenLynx. The green dots indicate that the target product ions were "found" by mass spectral detection and red dots represents "not found." This OpenLynx representation of LC/UV/ELSD/MS data does not show the purity

LC/MS Analysis of Library Validation Plates

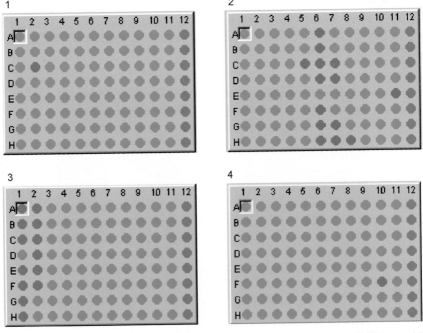

Figure 2.5 Plateview of the analytical data for validation plates of a 3D library with biphenyl scaffold.

assessment. However, it obviously displays that the 17th and the 18th primary/ secondary amines (columns 6 and 7 of plate 2) and the 23rd and the 24th (columns 1 and 2 of plate 3) primary/secondary amines failed to produce desired compounds. These four reagents were replaced with substitutes and the library production step was carried on.

7 LIBRARY PRODUCTION STAGE

The library production phase actually consisted of three elements, library synthesis, in-process QC, and support cleavage. For the library synthesis, one manipulated the cans by performing the chemistry as validated. Typical glassware was used and cans were washed by use of a large separatory funnel for each step. The ability to use typical glassware was a significant advantage of the Irori system as one only had to perform the synthetic operation one time in order to make multiple compounds.

In-process QC was performed through the use of tracing cans. A tracing can is defined as a can, which is visibly identifiable, contains the solid support and undergoes the same reaction processes as the library, thereby upon cleavage providing reaction products. The reaction products were analyzed in order to obtain analytical information and to "trace" library synthetic processes. We used MiniKans (Fig. 2.4) as tracing cans that were larger than MicroKans. These cans were intermixed throughout the library and were identifiable by the naked eye because of their size difference. These tracing cans contained an RF tag and were identified using the AccuTag reader. These compounds were cleaved from the solid support and submitted without purification to mimic the actual cleavage conditions. The purpose was to elucidate any difficulties in the library production process. If any additional time was need for reaction completion a new tracing can would be removed and checked for completion. During the library production process, samples of in-process compounds could be submitted for analytical review without removing any actual library members by means of tracing cans.

The final phase of library production was compound cleavage. The individual library compounds were contained on the solid support and must be cleaved to determine analytical purity and be prepared for submission. One may cleave in an arbitrary array by placing the compounds in a cleavage block randomly and adding cleavage solution, however, in our paradigm we must cleave each compound into a specific location on a 96-well plate. In our laboratory we have modified the AccuSort 10 K to handle sorting into Bohdan MiniBlocks as opposed to using Irori's Accu-cleave technology. The modifications involved developing a deck to hold the MiniBlocks, and generation of an algorithm to place the cans into the desired locations. Figure 2.6a shows the AccuSort 10 K and Fig. 2.6b exhibits detailed view of AccuSort 10 K operating with a Bohdan MiniBlock.

After sorting of the compounds into the desired locations the appropriate cleavage solution was added and the blocks were drained into the 96-well capture plate. These solutions were then concentrated for determination of analytical purity, typically in acetonitrile or methanol. A small aliquot of this solution was submitted

Figure 2.6   (a) Picture of AutoSort 10 K; (b) close-up view of
AutoSort 10 K in operation with Bohdan MiniBlock.

for analytical QC as a "daughter" plate. The remainder was transferred to a pre-
weighed and tared 2D-barcoded tube for final drying and weight determination and
for compound submission.

8   HIGH THROUGHPUT PURIFICATION USING SPE

Chromatographic purification of the library compounds requires a medium to high
throughput system and there have been reports of purification based on UV or MS

in the literature.[18,24] As a way of rapid semi-purification without going through individual chromatographic purification, automated solid-phase extraction was employed in our group for most of the libraries produced. Since individual extraction methods cannot be developed for each and every member of the library, a general feature of the library synthesis was utilized for purification. An example shown in Fig. 2.7 exploited the ionic interactions between SPE sorbents and impurities in the library. The product mixture contained pyridine and a positively charged pyridinium species in addition to the desired indolizine product. Because the impurities were either basic or positively charged while the desired products were neutral, strong cation exchange (SCX) SPE cartridges were used to purify the product from other contaminants. The samples in ethyl acetate were loaded onto the 96-well format SCX cartridges (Versaplate from Varian) and eluted with appropriate amounts of ethyl acetate. The neutral indolizine products were eluted through the cartridges and collected onto collection plates, while the pyridine and pyridinium species were trapped in the SCX cartridge *via* ion-exchange mechanism. Depending on the complexity of SPE methods, that typically included solvent adding and mixing, conditioning, loading, washing, and eluting steps, the run time ranged from half hour to two hours per 96-well plate. SLE using Hydromatrix (Varian) is also possible.[25]

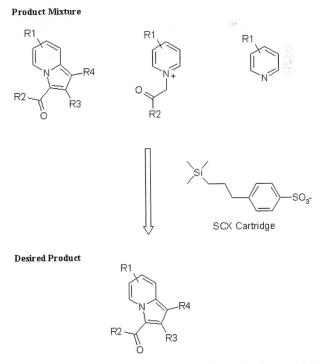

Figure 2.7   An example of SPE as a rapid purification method of combinatorial library.

9   100% QC BY LC/UV/ELSD/MS AND ANALYTICAL REPORTS

Each and every sample produced in the library was analyzed by LC/UV/ELSD/MS and processed automatically using OpenLynx software. The chemists can view the analytical data through OpenLynx browser on their PC, as shown in Fig. 2.8. The plate view in the upper left hand corner shows an overview of the analytical data by indicating wells "found" by MS detection with green dots and wells "not found" with red dots. The detailed data in the well location C02 are shown in this figure. For this specific sample, the target MW of 328 was detected by the mass spectrometer and the retention time of the peak was 1.45 min. The mass spectrum of this peak is shown in the middle. The UV and ELSD traces shown at the bottom

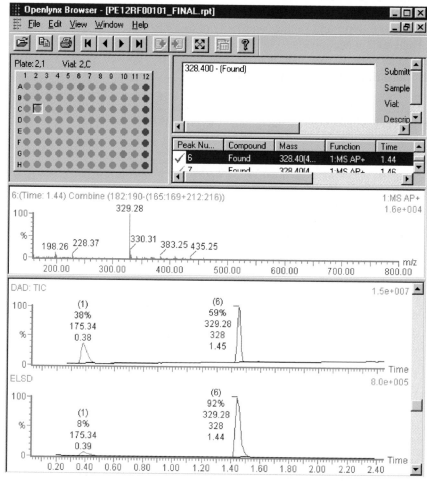

Figure 2.8   OpenLynx browser view of a library example.

indicate the purity of the sample, with an impurity at the retention time of 0.38 min. Samples in column 12 with blue dots were QC standard and blank samples for instrument performance verification. After reviewing the analytical data, chemists made the decision whether to submit the library plate for biological screening. Since we have not yet implemented the individual sample culling the decision was plate-based.

Once all the submission plates for the library were decided, the dried mother plate samples were weighed in pre-tared tubes for gravimetric quantitative analysis. Analytical reports written for each library included the description of the library, analytical procedures, statistics and graphs. Examples of the charts for the analytical reports are shown in Fig. 2.9 including: (a) the percentage of "MS found", (b) sample purity by UV and ELSD, and (c) the average weight per plate. In addition, a graphical view of the analytical data was generated using Spotfire software (Spotfire, Somerville, MA) and it was included in the analytical report in a picture format. In Fig. 2.10, a library example of this Spotfire view is shown. From left to right, the structure of the sample (not the actual structure of this sample), the graphical view of the analytical data in 96-well plate format, and detailed associated information are displayed. The information shown belongs to the sample in G02 of the plate 0108RF005. In the graphic representation section, circle indicated samples that were found by mass spectral detection and X mark indicated samples that were not found. The color (from red to green) and the size (from small to large) of the circle varied with the purity of the sample in order to give a quick overview of the library quality.

The last step of the analytical data handling was to import the data into a database for future viewing and querying. A customized database was developed in collaboration with Groton NeoChem (Acton, MA, www.neochem.com). The database stored the actual LC and MS data in addition to the metadata. A library example imported into this database is shown in Fig. 2.11. From the top left section and clockwise are displayed 96-well plate view, compound structure (not the actual structure of this sample) and associated information, a mass spectrum of the chosen peak, and LC chromatograms. This database is capable of importing NMR data as well, but we do not yet routinely run complete NMR analysis for combinatorial libraries.

10   FUTURE TRENDS

The incorporation of individual sample culling capability is in progress using individual 2D-barcoded sample tubes. A library compound will not be identified by the plate ID and well location any more when this conversion is complete; instead it will be identified by its own barcode. As a result, reformatting after a culling step would be much easier and simpler. As the pursuit for the never-before-synthesized compounds leads to novel and possibly more complex library synthetic steps, optimization of synthesis alone cannot achieve the high purity desired. Therefore high throughput chromatographic purification is actively explored. In this regard,

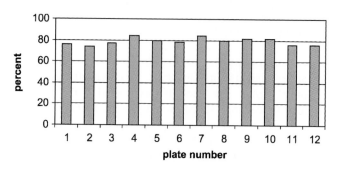

a

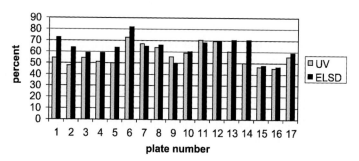

b

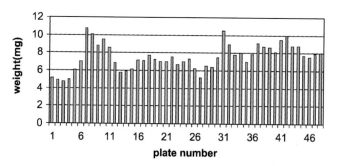

c

Figure 2.9 Graphs included in an analytical report for library: (a) the percentage of MS found per plate; (b) UV/ELSD purity per plate; (c) the average weight per plate.

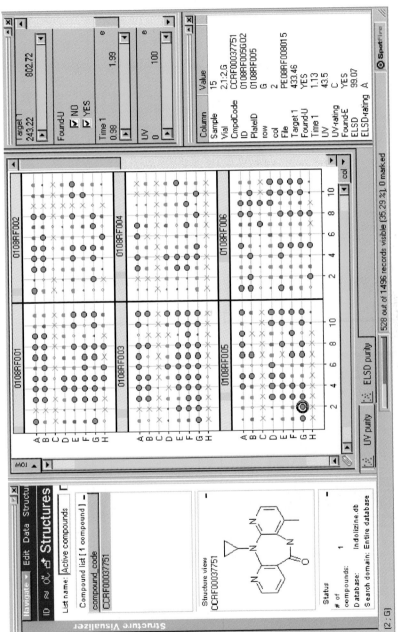

Figure 2.10 Spotfire view of a library example.

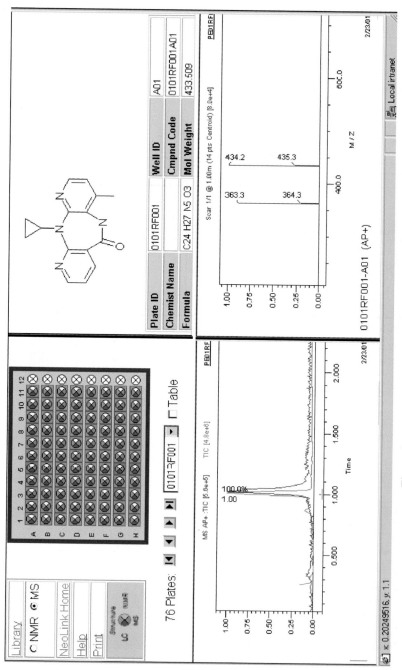

Figure 2.11 NeoChem database view of a library example.

SFC is gradually being embraced as a routine method to take advantage of its high speed and non-requirement for solvent evaporation step after fraction collection. Applicability of chiral separation with SFC is also an appeal of this technique.

11 CONCLUSION

We reported the synthetic and analytical procedures employed in our group using split-and-pool Irori MicroKan method. This method takes advantage of not only split/pool synthesis on solid support but also parallel approach producing one sample per well. Through automation and multiplexing 100% QC by LC/UV/ELSD/MS analysis of all the library compounds produced was achieved. Employing efficient and effective software was essential for the massive analytical data processing and presentation. New technologies to enhance the purity and quality of the library including culling and reformatting procedures, high throughput chromatographic purification, and SFC are continuously explored.

ACKNOWLEDGEMENTS

We thank Gregory Roth for his critical reading of the manuscript. Authors gratefully acknowledge George Lee, Scott Leonard, Daniel Marshall, Gustavo Rodriguez, Pingrong Liu, John Lord, Lida Zadeh and Noel Wilson for their valuable input.

REFERENCES

1. C.N. Selway and N.K. Terrett, *Bioorg. Med. Chem.*, **4**, 645 (1996).
2. J.A. Ellman and L.A. Thompson, *Chem. Rev.*, **96**, 555 (1996).
3. E.M. Gordon, R.W. Barrett, W.J. Dower, S.P.A. Fodor, and M.A. Gallop, *J. Med. Chem.*, **37**, 1385 (1994).
4. J.J. Baldwin, J.J. Burbaum, I. Henderson, and M.H.J. Ohlmeyer, *J. Am. Chem. Soc.*, **117**, 5588 (1995).
5. A. Furka and W.D. Bennett, *Comb. High Throughput Screen.*, **2**, 105 (1999).
6. A.R. Vaino and K.D. Janda, *Proc. Natl Acad. Sci. USA*, **97**, 7692 (2000).
7. J.W. Guiles, C.L. Lanter, and R.A. Rivero, *Angew. Chem. Int. Ed. Engl.*, **37**, 927 (1998).
8. C.L. Lanter, J.W. Guiles, and R.A. Rivero, *Mol. Div.*, **4**, 149 (1999).
9. E.J. Moran, S. Sarshar, J.F. Cargill, M.M. Shahbaz, A. Lio, A.M.M. Mjalli, and R.W. Armstrong, *J. Am. Chem. Soc.*, **117**, 10787 (1995).
10. K.C. Nicolaou, X.-Y. Xiao, Z. Parandoosh, A. Senyei, and M.P. Nova, *Angew. Chem. Int. Ed. Engl.*, **34** (20), 2289 (1995).
11. K.L. Morand, T.M. Burt, B.T. Regg, and T.L. Chester, *Anal. Chem.*, **73**, 247 (2001).
12. J.N. Kyranos and J.C. Hogan, Jr., *Anal. Chem.*, **70**, 389A (1998).
13. T. Wang, L. Zeng, T. Strader, L. Burton, and D.B. Kassel, *Rapid Commun. Mass Spectrom.*, **12**, 1123 (1998).
14. W. Goetzinger and J.N. Kyranos, *Am. Lab.*, **30**, 27 (1998).

15. H. Lee, L. Li, and J.N. Kyranos, *Proc. 47th ASMS Conf.*, 2715 (1999).
16. J.B. Li and J. Morawski, *LC–GC*, **16**, 468 (1998).
17. J.P. Kipliger, R.O. Cole, S. Robinson, E.J. Roskamp, R.S. Ware, H.J. O'Connell, A. Brailsford, and J. Batt, *Rapid Commun. Mass Spectrom.*, **12**, 658 (1998).
18. L. Zeng and D.B. Kassel, *Anal. Chem.*, **70**, 4380 (1998).
19. M.C. Ventura, W.P. Farrell, C.M. Aurigemma, and M.J. Greig, *Anal. Chem.*, **71**, 4223 (1999).
20. I. Hughes and D. Hunter, *Curr. Opin. Chem. Biol.*, **5**, 243 (2001).
21. J.N. Kyranos, H. Cai, D. Wei, and W.K. Goetzinger, *Curr. Opin. Biotechnol.*, **12**, 105 (2001).
22. D.B. Kassel, *Chem. Rev.*, **101**, 255 (2001).
23. S.V. Ley, I.R. Baxendale, R.N. Bream, P.S. Jackson, A.G. Leach, D.A. Longbottom, M. Nesi, J.S. Scott, R.I. Storer, and S.J. Taylor, *J. Chem. Soc., Perkin Trans.*, **1**, 3815 (2000).
24. H.N. Weller, *Mol. Div.*, **4**, 47 (1998).
25. S.X. Peng, C. Henson, M.J. Strojnowski, A. Golebiowski, and S.R. Klopfenstein, *Anal. Chem.*, **72**, 261 (2000).
26. J.G. Breitenbucher, K.L. Arienti, and K.J. McClure, *J. Comb. Chem.*, **3**, 528 (2001).
27. L. Yang, T.D. Mann, D. Little, N. Wu, R.P. Clement, and P.J. Rudewicz, *J. Anal. Chem.*, **73**, 1740 (2001).
28. C.E. Kibbey, *Mol. Div.*, **1**, 247 (1995).
29. L. Fang, M. Wan, M. Pennacchio, and J. Pan, *J. Comb. Chem.*, **2**, 254 (2000).
30. A. Stolyhwo, H. Colin, M. Martin, and G.J. Guiochon, *J. Chromatogr.*, **288**, 253 (1984).
31. G. Guiochon, A. Moysan, and C. Holley, *J. Liq. Chromatogr.*, **11**, 2547 (1988).
32. E.M. Thurman, I. Ferrer, and D. Barcelo, *Anal. Chem.*, **73**, 5441 (2001).

3

High Throughput Flow Injection Analysis–Mass Spectrometry

Kenneth L. Morand

Procter & Gamble Pharmaceuticals, Inc., Health Care Research Center,
8700 Mason-Montgomery Road, OH 45040, USA

CONTENTS

1 INTRODUCTION

Many of the existing synthetic techniques, as well as the associated analytical quality control (QC) methods, for solution-phase combinatorial chemistry have their origins in practices used extensively in classical medicinal and analytical chemistry. A critical difference and challenge, nevertheless, is the need to translate these practices, *i.e.* synthesis and characterization, into viable procedures that effect the production of large numbers of drug-like compounds. In analytical chemistry,

High Throughput Analysis for Early Drug Discovery
Edited by James N. Kyranos

an example of one such difference between traditional and combinatorial synthesis can be connected to the increased importance and need of mass spectrometry (MS) in compound array and new chemical entity (NCE) QC. As is now typified across the pharmaceutical industry, the requirement for MS characterization and molecular weight (MW) determination has increased nearly 100-fold in the previous decade (Fig. 3.1) with the growth of combinatorial and automated chemistry.

The present chapter, therefore, addresses the technology and application of high throughput flow injection mass spectrometry (FIA–MS) for QC and synthesis verification for spatially addressable libraries. In spite of the advances in high-speed HPLC/MS, flow injection analysis, either in combination with MS[1] or other detection systems,[2] is still utilized extensively to support combinatorial chemistry and related drug discovery applications. The high sampling rates afforded by FIA analysis, in the absence of chromatographic separation, make FIA–MS ideally suited for those applications requiring QC analysis or MW verification of large compound arrays and mixtures. In general, technology advances with instrumentation and sample handling procedures have led to significant reductions in the sample analysis times for FIA–MS resulting in greater than an order of magnitude increase in MS instrumental sample capacity (Fig. 3.1). Further, high-speed FIA–MS instrumentation has, necessarily, been combined with the use of intelligent data acquisition and interpretation software packages to facilitate rapid QC analysis of compound arrays using automated sample MW identification and confirmation tools.[3-6] To this end, high-speed FIA–MS analysis for use in combinatorial chemistry has been well established in the laboratory. This chapter is provided, therefore, for the potential newcomer to combinatorial chemistry or high-speed characterization and offers details of the advantages and issues in using flow injection analysis for the characterization of large, spatially addressable compound arrays.

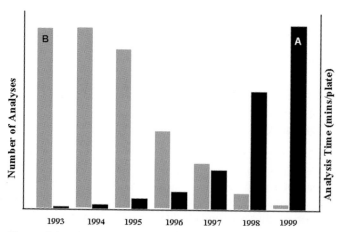

Figure 3.1 Trends in (a) the annual number of sample analyses performed for new chemical entity characterization and (b) the reduction in the total analysis time for a single sample using FIA–MS.

2   OPEN-ACCESS MASS SPECTROMETRY

For the analytical chemist, no single analytical instrument, technique or application is adequate to address the many and varied needs in high throughput medicinal chemistry. Organic compound libraries are synthesized in solution, using soluble polymer supports or on solid-supports. They may be enumerated as a collection of single chemical entities or as a single collection of a number of chemical entities. Similarly, they may be synthesized for the purpose of building diversity within a corporate collection or within a theme with the intention of generating structure activity relationships (SAR) against a given biological target. Nevertheless, each approach has its own advantages, regardless of the specific synthetic format, array size, or use, that must be considered in order to make certain the proper qualification of the compound library.[7–9]

There have been numerous analytical and MS applications developed for high throughput medicinal chemistry. Many of the more established techniques are addressed in this or other books germane to the topic.[10,11] However, typically overlooked among the high profile developments of high-speed chromatography,[12–15] solid phase[16] and flow NMR[17] and mass spectrometry-based purification techniques,[18,19] the introduction of open-access or walk-up MS (OAMS) can be regarded as the foundation of high-speed MS-based applications for combinatorial chemistry. Pullen[20–22] and Taylor[23] introduced the concept of OAMS in a series of manuscripts in the mid-1990s. Initially developed as a method to provide the medicinal chemist direct access to analytical QC technologies, the introduction of OAMS, nevertheless, laid the groundwork for many of the hardware and software applications that are now routinely used to support the combinatorial chemistry. Since its introduction, OAMS has proven to be an invaluable tool for medicinal chemistry. Rapid MW determinations using open-access FIA–MS and, more recently, reaction monitoring using open-access HPLC–MS have significantly reduced the time necessary for developing new chemistry schemes. More importantly, OAMS has provided the medicinal chemist with tools that afford the direct and immediate feedback necessary for synthesis optimization.

3   HIGH-SPEED FLOW INJECTION ANALYSIS–MASS SPECTROMETRY

High-speed FIA–MS was developed as a natural extension of open-access MS to satisfy a need for reliable and high-speed QC methods to validate combinatorial chemistry production. It should be noted that there are a great number of analytical instrumentations and applications that have been successfully developed and which are currently being used to support combinatorial chemistry QC analysis.[10,11] However, without question mass spectrometry and FIA–MS have resulted in the greatest impact on increasing the throughout and speed at which samples may be characterized to benefit combinatorial chemistry. A principal advantage of

FIA–MS resides in the ability of the analyst to readily calculate the MW of the expected synthetic product and, subsequently, to directly observe and measure its calculated mass. The advent of soft atmospheric pressure ionization techniques for MS, such as atmospheric pressure chemical ionization (APCI) and electrospray ionization (ESI) which predominately result in the generation of a charged species representative of the product compound, have made this process all the more straightforward for the non-expert user by largely eliminating the need for detailed data interpretation or even expert user intervention. In contrast to the cumbersome data interpretation requirements customary when using electron impact ionization (EI), flow injection analysis in combination with ESI or APCI MS by and large provides a direct measurement of the product compound MW. Again, combining the readied MW measurement capabilities of FIA with automated data reduction routines have allowed the analyst to rapidly analyze hundreds to thousands of samples from combinatorial chemistry in a fraction of the time required using other analytical characterization approaches.

A typical high throughput or high-speed FIA–MS system is illustrated in Fig. 3.2. In general, the system is simple in design consisting of a flatbed liquid handling robot or other autosampling device capable for sampling from a microtiter plate, an injection module, a suitable pumping system, and mass spectrometer. The liquid handling robot and injection module, when combined, function as the system autosampler and interface to the mass spectrometer. A typical autosampler of this type is the Gilson 215 liquid handling robot (Gilson, Inc., Madison, WI, USA) with a Gilson 819 single port injection module, however, suitable autosamplers are available from all major HPLC vendors. Because the present discussion is centered on the use of FIA–MS for QC analysis of combinatorial chemistry arrays, a common issue is the rate of analysis or sample throughput for a given analysis technique. Analysis cycle time for a system of this basic design, and as delivered by the manufacturer, are on the order of 20–30 s per sample or approximately 30–45 min per 96-well microtiter plate. This is certainly not high-speed analysis by current standards within the industry, but, nevertheless, the standard single-probe or single-injector FIA–MS instrument is well suited for most applications that require MW determination or ID, and they allow notably faster sample characterization than HPLC/MS analysis. There are, however, numerous ways by which the analyst

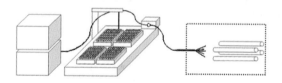

Pump   Autosampler   Mass Spectrometer

Figure 3.2   General instrument schematic for FIA–MS system including an HPLC pump, flat-bed liquid handling robot and injection module and single quadruple mass spectrometer.

can markedly increase the analysis speed and improve the overall sample throughput for FIA–MS. The following sections review the approaches taken by a number of researchers in the pharmaceutical industry to increase the speed of analysis or, alternatively, sample throughput. Fortunately, this is a straightforward process for FIA–MS as there but a limited number of instrumental variables to optimize, *i.e.* mobile phase flow rate, autosampler injection cycle, dead volume and the supply of compounds to and from the autosampler. Each of these issues will be addressed in succession.

3.1   Flow Rate, Injection Loops, and Transfer Tubes

Flow injection analysis involves the direct transfer of a soluble plug of the analyte by means of a liquid flow to the mass spectrometer ion source. Because there is no chromatographic retention of the analyte, the length of time for the analysis is directly related to time required to transfer the flow injection plug from the injection loop to the source of the mass spectrometer with exception to band broadening that will occur due to the internal volumes of the injection loop and transfer tubing. For FIA–MS analysis the sample analysis speed is limited, for the most part, not by the speed of the autosampler injector, but instead by the width of the flow injection plug as observed at the ion source of the mass spectrometer. The rate of the autosampler injection cycle is only a limitation to the overall speed of analysis to the degree by which the analyst has taken to minimize the peak widths, since any amount of optimization or elimination of lengthy injector cycle functions will not overcome poor plumbing and excess connector dead volumes.

To address the issue of peak broadening, the flow injection plug or peak can be approximated as a normal distribution of signal against time with its temporal width described by the standard deviation of the distribution. For FIA, it is necessary to consider only a few contributors to peak width: (1) the initial width of the analyte plug due to the sample loop volume, (2) transport of the plug from the sample loop, (3) transport through the tube connecting the injector to a splitter, and (4) transport through the tube connecting the splitter to the mass spectrometer ion source. Further, flow splitting prior to the mass spectrometer, and the problems introduced by flow splitting, may be ignored in many FIA–MS applications if the total mobile phase flow rate is sufficiently small enough such that the entire flow can be directed to the ion source of the mass spectrometer.

Initially, the flow injection plug exists in a uniform distribution, evenly distributed over the length of the filled sample loop. The standard deviation, σ, of a uniform distribution is simply $1/\sqrt{12}$ times its total width. Therefore, the standard deviation, in time units, caused by the initial sample volume is

$$\sigma_s = \frac{1}{\sqrt{12}} \frac{V_s}{F} \qquad (3.1)$$

where V_s is the loop volume and F is the flow rate through the loop. Figure 3.3 has been adapted from Dicesare[24] and depicts the general trend of increasing the loop volume, or partial loop fill, and the subsequent effect of increasing the standard

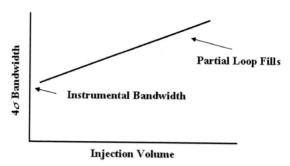

Figure 3.3 General trend of increasing the loop volume, or partial loop fill, and the subsequent effect of increasing peak width standard deviation. (Adapted from J.C. DiCesure, M.W. Wong, and J.G. Alwood, *J. Chromatogr.*, **217**, 369–386 (1981).)

deviation of the injection peak width. However, one may extend Eq. (3.1) for the short, open transfer tubes associated with FIA and approximate the entire flow path from the injection loop to the mass spectrometer as the sample injection region. In this case, the total standard deviation of the peak as observed by the mass spectrometer may be approximated, therefore, by the summation of the squares of the standard deviations for each individual component in the injection loop region.

$$\sigma^2\text{total} = \sigma^2\text{loop} + \sigma^2\text{transfer tubes} + \sigma^2\text{internal volumes}$$

A common approach used to increase FIA–MS sample analysis speed has been to reduce the standard deviation of the flow injection plug such that a higher density, or larger number, of flow injection plugs may be observed in a given time window without sacrificing peak resolution. Reduction of the peak width can be accomplished by either minimizing the internal volume of the injection loop and transfer tubes (contribution due to the internal volumes of the valves and fittings are considered to be negligible) or by increasing the mobile phase flow rate through the injection region (Eq. (3.1)). In practice, a combination of these approaches is used to increase analysis speed within the constraints of the instrument design, *e.g.* mobile phase backpressure and sampling speed of the autosampler. Nevertheless, sample analysis times are reduced from 120 to 180 s per sample for standard OAMS instrumentation to approximately 30 s per sample.

3.2 Autosampler Overhead – Multiplexing Injection Cycle Procedures

To this point in our discussion we have ignored the contribution of the autosampler, and specifically the injection cycle, to the overall speed and throughput of FIA–MS, instead addressing methods to increase the transfer rate of the flow injection plug to the mass spectrometer. Nevertheless, the overhead associated with injection cycle must be taken into account in order to fully realize the potential of high-speed FIA. Sampling, injector loading, probe washing, *etc.*, all potentially lengthen the time on the autosampler and contribute to limit the overall sample analysis rate. One approach that has been employed to minimize the lengthy injection cycle routines incorporates a multiprobe autosampler with the capability to load multiple injection loops simultaneously.

Wang *et al.*[25] first reported on this approach by interfacing a Gilson 215 multiprobe liquid handling robot and eight-position injection module to a PE Sciex single quadrupole mass spectrometer. In their design, the liquid handling robot and multiplexed injection module allows eight samples to be simultaneously loaded into individual injection valves. The injection valves are triggered sequentially such that each sample is analyzed in a separate time window offset by the delay time between triggering each injection valve (Fig. 3.4). Because the samples are loaded onto all eight-injection valves simultaneously, a marked improvement in the batch analysis time is realized by eliminating a significant fraction of the autosampler routines,

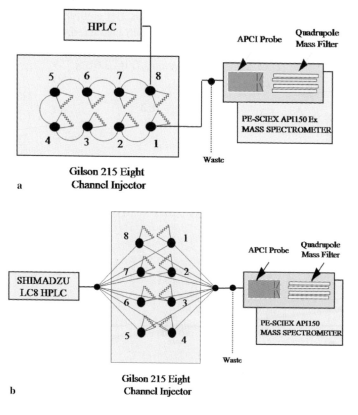

Figure 3.4 Connections and flow path for 8 injectors configured in (a) a serial mode of operations and (b) a parallel mode of operation. (Adapted from T. Wang, L. Zeng, T. Strader, L. Burton, and D. Kassel, *Rapid Commun. Mass Spectrom.*, **12**, 1123–1129 (1998).)

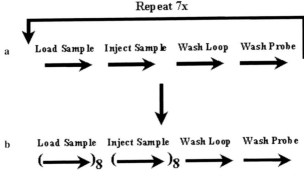

Figure 3.5   Comparison of the injection cycle sequence to the analysis of eight samples using (a) a standard single probe autosampler and (b) an eight-probe autosampler.

i.e. probe washing and sample loading (Fig. 3.5). For example, the authors reported analysis times of approximately 45 s for a single set of eight injections, and a total analysis time on the order of 10 min for FIA–MS characterization of a complete 96-well microtiter plate. This is nearly a 5-fold increase in analysis speed over conventional single probe autosampler systems (Fig. 3.6).

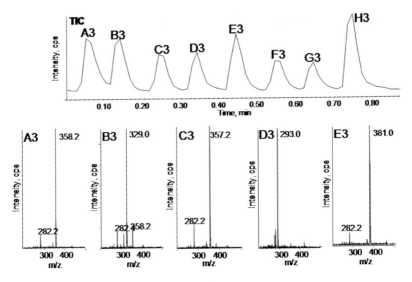

Figure 3.6   Analysis of 96-well microliter plate using a serial injection configuration. TIC chromatogram shows that samples A3–H3 are baseline separated and the FIA–MS spectra for each of the samples is accomplished in less than 0.8 min (<48 s). ESI mass spectra for the first 5 samples are shown. No peak tailing was observed and consequently, no carry-over was observed from one sample to the next. (Adapted from T. Wang, L. Zeng, T. Strader, L. Burton, and D. Kassel, *Rapid Commun. Mass Spectrom.*, **12**, 1123–1129 (1998).)

3.3 Combining High Sampling Rates with Fast Analysis

The use of multiprobe autosamplers has significantly impacted sample throughput and the design for both flow injection and liquid chromatography interfaces to MS.[26,27] However, the improvements in the sampling rate achieved with multiprobe systems are not without their drawbacks as these gains are made generally made at the expense of other system parameters. With most marketed multi-valve injection systems it is necessary to split the total solvent flow amongst the separate sample injectors, hence, the flow rate through each injector is a fraction of the total system flow rate. If the total flow rate through an eight-valve injection module is 5 ml/min, then the flow rate through each injector is only 625 μl/min (Fig. 3.7). Efforts to minimize the overhead associated with the injection cycle of the autosampler are accomplished with a marked decrease in the flow rate through the injection module. In order to improve the overall sample analysis rate, it is necessary to increase the linear velocity, or transfer rate, of the flow injection plug through the injection module. This may be accomplished by fabricating an autosampler that incorporates the efficiencies of the multiprobe design for improving sampling rates with the high linear velocities and narrow peak widths associated with single probe autosamplers. A schematic diagram of the injection module for an autosampler with this configuration is shown in Fig. 3.8.

Morand *et al.* have reported the design and application of an autosampler of this design to benefit high-speed FIA–MS QC analysis[28] and, separately, to support very rapid bioanalytical quantitation using supercritical fluid chromatography (SFC)–MS.[29] In this design, each eight-way flow splitter, which splits and recombines the mobile phase flow prior to and down stream from the multiple injection ports (refer to Fig. 3.4), was replaced with a set of actuated eight-position flow path selection valves. By decoupling or isolating the multiple injection ports from one another using the column selection valves the mobile phase flow may be directed to each injection port sequentially and without reducing the total flow rate through all ports due to flow splitting. Using this design flow injection peak widths of approximately 0.5 s (FWHM) were achieved at the outlet and an injection sequence time of approximately 6.6 s for a single eight-sample loading (Fig. 3.9). Additionally, sample analysis times were approximately 5 min for a single 96-well microtiter plate (Fig. 3.10). It should be noted, however, due to the very high analysis speeds and

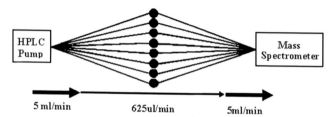

Figure 3.7   Representative diagram showing the total flow reduction through the injection region of a standard eight-port injection module due to flow splitting before and after the injection ports.

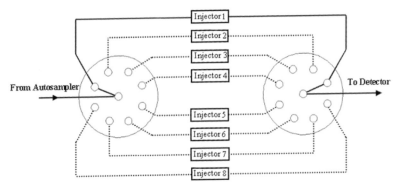

Figure 3.8 Flow diagram for the injection region of a modified eight-port injection module. The flow splitters before and after the injection ports have been replaced with two eight-position column selection valves. The solid line represents the "selected" flow path, while the dotted lines represent the remaining seven "isolated" flow paths. (Adapted from K. Morand, T. Burt, B. Regg, and T. Chester, *Anal. Chem.*, **73**(2), 247–252 (2001).)

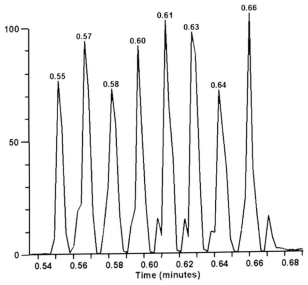

Figure 3.9 Single ion chromatogram of quinine (MW = 324.2 Da, $C_{20}H_{24}N_2O_2$), for a single set of 8 FIA–MS using a modified multiprobe autosampler incorporating column switching valves. The switching valves were triggered at a rate of ~0.8 s per sample. The peak widths were measured as 0.5 s at FWHM and 0.9 s at the base. (Reprinted with permission from K. Morand, T. Burt, B. Regg, and T. Chester, *Anal. Chem.*, **73**(2), 247–252 (2001).)

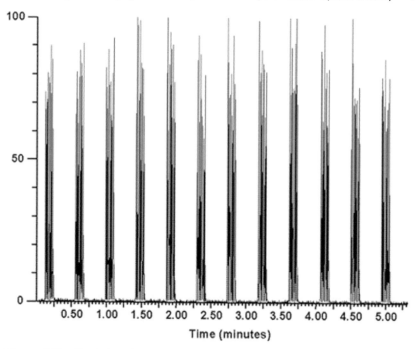

Figure 3.10 Single ion chromatogram of quinine (MW = 324.2 Da, $C_{20}H_{24}N_2O_2$), for a full 96-well microtiter plate using a modified multiprobe autosampler incorporating column switching valves. The entire plate was analyzed in approximately 5 min. The "dead-time" between each set of 8 injections is approximately 23 s, while the sample analysis time for each set of 8 injections is approximately 6.6 s. (Reprinted with permission from K. Morand, T. Burt, B. Regg, and T. Chester, *Anal. Chem.*, **73**(2), 247–252 (2001).)

minimum peak widths achieved with this design it was not practical to couple the autosampler with standard single quadrupole instrumentation. The slow scan rates of standard quadrupole instrumentation, *e.g.* 1 kDa/s, make such systems impractical for the analysis of flow injection peaks on the order of 0.5 s in width, when full scan data, or wide scan ranges, are required. Therefore, in order to obtain a sufficient number of sampling points (spectra) across each flow injection peak, the autosampler was coupled to atmospheric pressure ionization time-of-flight mass spectrometry for which the scan speed is on the order of 10–20 kDa/s. However, new quadrupole instrumentation has been recently introduced with a demonstrated scan speed of 5 kDa/s and could be used successfully at these sampling and analysis rates.

4 OTHER HIGH SPEED AND HIGH THROUGHPUT INSTRUMENTATION FOR MW MEASUREMENT

In concluding this chapter, it is worth noting that there are many direct analysis techniques other than FIA–MS that may be used for MW identification and QC analysis of large, spatially addressable compound arrays. Techniques, such as direct infusion and desorption ionization, were used extensively in the earliest developments of analytical methods for combinatorial peptide chemistry.[30-32] The emergence of microfluidic and microfabrication technologies, which offer a relatively simple design and small size applicable for the analysis of compound array formats, has resulted in the renewed development of these approaches as practical alternatives to FIA–MS.

Karger and coworkers[33] have demonstrated the utility of direct sampling from a microtiter plate using a subatmospheric electrospray interface eliminating the need for a separate pumping module (Fig. 3.11). The microtiter plate is held in a vertical position on three x, y, and z-stage stepper motors. The motors allow the microtiter plate to be positioned directly in front of a short, stationary sampling capillary, which is integrated into a miniaturized ESI interface and subatmospheric pressure chamber. Sample flow through the capillary is regulated by the pressure differential between the inlet and outlet of the capillary needle. Reducing the pressure in the subatmospheric chamber from nominally 1 atm effects flow of the sample to the ionization source. A set of two pressure control valves allow for sample rates from 200 to 100 nl/min. While sample analysis times are still short of the ultrafast analysis times achievable using FIA–MS, the low flow rates, approximately 200 nl/min, and minimal sample consumption allow ready access to tandem mass spectrometry experiments, which would otherwise not be practical using standard FIA–MS instrumentation (Fig. 3.12). Similar systems have been fabricated to incorporate parallel sampling[34] and micro-capillary electrophoresis.[35]

Finally, where direct sampling techniques simplify instrumental design by removing the autosampler and pumping module, desorption ionization techniques wholly remove any requisite for solvent flow introduction to the mass spectrometer. Matrix-assisted laser desorption has been used extensively

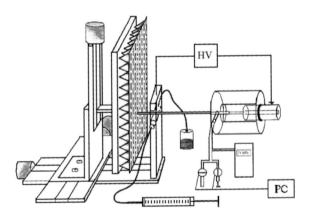

Figure 3.11   Diagram of a high throughput microtiter plate direct infusion system. (Reprinted with permission from C. Felten, F. Foret, M. Minarik, W. Goetzinger, and B. Karger, *Anal. Chem.*, **73** 1449–1454, (2001).)

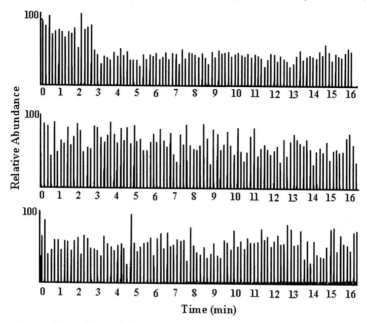

Figure 3.12   Direct infusion analysis of all wells from a 96-well microtiter plate containing 25 μl of angiotensin (0.1 mg/ml). The three traces represent a total of 288 direct infusion experiments completed in 48 min. The flow rate was approximately 1000 nl/min during the sample analysis. (Reprinted with permission from C. Felten, F. Foret, M. Minarik, W. Goetzinger, and B. Karger, *Anal. Chem.*, **73**, 1449–1454 (2001).)

for either the direct or indirect[31,36-38] characterization of bead-bound or split-pool libraries. However, the low mass matrix background associated with MALDI-TOF analysis makes it impractical for the QC analysis of combinatorial libraries. Desorption ionization on silicon (DIOS-MS) simplifies the use matrix-assisted laser desorption mass spectrometry (MALDI-MS) for small molecule analysis by using porous silicon to trap the analyte on a matrix free surface.[39,40] The use of the matrix-free surface allows small molecule analysis (<500 Da) with little or no fragmentation and without matrix interference in the lower MW range. Further, the silicon chips are inexpensive to manufacture and can be photo-patterned as an aid for visualization or to obtain mass spectra of multiple samples. For example, the surface may be etched in the pattern of a standard microtiter plate format to facilitate rapid sample spotting analysis using standard robotic equipment. With modest modifications to the sampling plates the chips are compatible with existing MALDI-TOF mass spectrometers. Sampling rates with current instrumentation are already on the order of commercial FIA–MS systems. It is expected that the evolution of laser technology will provide unique opportunities for applying DIOS MS for QC analysis of combinatorial chemistry libraries.

5 CONCLUSION

High-speed QC analysis has been of interest and concern since the introduction of combinatorial chemistry and high throughput medicinal chemistry. Consequently, flow injection analysis, direct infusion and desorption ionization MS techniques were established to satisfy the growing need for reliable and high-speed tools for the direct verification of the large numbers of compounds produced using the chemistry platforms. This chapter has outlined a few of the approaches that have been utilized to increase the overall sample analysis speed for MW determination. In light of these efforts, analysis rates have been markedly increased since the initial introduction of automated and OAMS.

ACKNOWLEDGEMENTS

The author would like to acknowledge the contributions and collaborations of Tom Burt, Tom Chester, Brian Regg, Debra Tirey and Roy Dobson for their help in the preparation of this chapter.

REFERENCES

1. D.B. Kassel, *Chem. Rev.*, **101**, 255–267 (2001).
2. J.N. Kyranos, H. Cai, B. Zhang, and W.K. Goetzinger, *Curr. Opin. Drug Discov. Dev.*, **4**(6), 719–728 (2001).
3. G. Hegy, E. Goerlach, R. Richmond, and F. Bitsch, *Rapid Commun. Mass Spectrom.*, **10**, 1894–1900 (1996).
4. R. Richmond, E. Goerlach, and J.-M. Seifert, *J. Chromatogr. A*, **835**, 29–39 (1999).
5. H. Tong, D. Bell, K. Tabei, and M.M. Siegel, *J. Am. Soc. Mass Spectrom.*, **10**, 1174–1187 (1999).
6. R. Richmond and E. Goerlach, *Anal. Chim. Acta*, **390**, 175–183 (1999).
7. S. Miertus and G. Fassina, *Combinatorial Chemistry and Technology: Principles, Methods and Applications*, Marcel Dekker, New York (1999).
8. B.A. Bunin, *The Combinatorial Index*, Academic Press, San Diego (1998).
9. S.R. Wilson and A.W. Czarnik, *Combinatorial Chemistry: Synthesis and Application*, Wiley, New York (1997).
10. B. Yan, *Analytical Methods in Combinatorial Chemistry*, Technomic Publishing Co., Lancaster (2000).
11. M.E. Schwartz, *Analytical Techniques in Combinatorial Chemistry*, Marcel Dekker, New York (2000).
12. I.M. Mutton, *Chromatographia*, **47**(708), 1–8 (1998).
13. W.K. Goetzinger and J.N. Kyranos, *Am. Lab.*, **30**(8), 27–37 (1998).
14. K. Yu and M. Balogh, *LC–GC*, **19**(1), 60–72 (2000).
15. Y. Cheng, Z. Lu, and U. Neue, *Rapid Commun. Mass Spectrom.*, **15**, 141–151 (2002).
16. M.J. Shapiro and J.S. Gounarides, *Biotechnol. Bioeng.*, **71**(2), 130–148 (2001).
17. P.A. Keifer, S.H. Smallcombe, E.H. Williams, K.E. Salomon, G. Mendez, J.L. Belletire, and C.D. Moore, *J. Comb. Chem.*, **2**, 151–171 (2000).

18. J.P. Kiplinger, R.G. Cole, S. Robinson, E. Roskamp, R.S. Ware, H.J. O'Connell, A. Brailsford, and J. Batt, *Rapid Commun. Mass Spectrom.*, **12**, 658–664 (1998).
19. L. Zeng, L. Burton, K. Yung, B. Shushan, and D.B. Kassel, *J. Chromatogr. A*, **794**, 3–13 (1998).
20. D.V. Brown, M. Dalton, F.S. Pullen, G.L. Perkins, and D. Richards, *Rapid Commun. Mass Spectrom.*, **8**, 632–636 (1994).
21. F.S. Pullen and D.S. Richards, *Rapid Commun. Mass Spectrom.*, **9**, 188–190 (1995).
22. F.S. Pullen, G.L. Perkins, K.I. Burton, R.S. Ware, M.S. Teague, and J.P. Kiplinger, *J. Am. Soc. Mass Spectrom.*, **6**, 394–399 (1995).
23. L.C.E. Taylor, R.L. Johnson, and R.J. Raso, *J. Am. Soc. Mass Spectrom.*, **6**, 387–393 (1995).
24. J.L. DiCesare, M.W. Dong, and J.G. Atwood, *J. Chromatogr.*, **217**, 369–386 (1981).
25. T. Wang, L. Zeng, T. Strader, L. Burton, and D. Kassel, *Rapid Commun. Mass Spectrom.*, **12**, 1123–1129 (1998).
26. A.B. Sage, D. Little, and K. Gilles, *LC–GC*, **18**, S20–S29 (2000).
27. T. Wang, J. Cohen, D.B. Kassel, and L. Zeng, *Comb. Chem. High Throughput Screen*, **2**, 327–334 (1999).
28. K.L. Morand, T.M. Burt, B.T. Regg, and T.L. Chester, *Anal. Chem.*, **73**(2), 247–252 (2001).
29. S.H. Hoke, J.A. Tomlinson, R.D. Bolden, K.L. Morand, J.D. Pinkston, and K.R. Wehmeyer, *Anal. Chem.*, **73**(13), 3083–3088 (2001).
30. B.J. Enger, G.J. Langly, and M. Bradly, *J. Org. Chem.*, **60**, 2652–2653 (1995).
31. C.L. Brummel, J.C. Vickermann, S.A. Carr, M.E. Hemling, G.D. Roberts, W. Johnson, J. Weinstock, D. Gaitanoloulos, S.J. Benkovic, and N. Winograd, *Anal. Chem.*, **68**, 237–242 (1996).
32. R.S. Youngquist, G.R. Fuentes, M.P. Lacey, and T. Keough, *J. Am. Chem. Soc.*, **117**, 3900–3906 (1995).
33. C. Felten, F. Foret, M. Minarik, W. Goetzinger, and B. Karger, *Anal. Chem.*, **73**, 1449–1454 (2001).
34. H. Liu, C. Felten, Q. Xue, B. Zhang, P. Jedrzejewski, B. Karger, and F. Foret, *Anal. Chem.*, **72**, 3303–3310 (2000).
35. B. Zhang, F. Foret, and B. Karger, *Anal. Chem.*, **73**, 2675–2681 (2001).
36. M.L. Pacholski and N. Winograd, *Chem. Rev.*, **99**, 2977–3005 (1999).
37. C. Enjalbal, D. Maux, J. Martinez, R. Combarieu, and J.-L. Aubagnac, *Comb. Chem. High Throughput Screen*, **4** (4), 363–373 (2001).
38. M.C. Fitzgerald, K. Harris, C.G. Shevlin, and G. Suizdak, *Bioorg. Med. Chem. Lett.*, **6** (8), 979–982 (1996).
39. J. Wei, J.M. Buriak, and G. Suizdak, *Nature*, **399**, 243–246 (1999).
40. Z. Shen, J.J. Thomas, C. Averbuj, K.M. Broo, M. Engelhard, J.E. Crowell, M.G. Finn, and G. Suizdak, *Anal. Chem.*, **73** (3), 612–619 (2001).

4

High Throughput Flow Injection Analysis–Mass Spectrometry for Combinatorial Chemistry Using Electrospray Ionization, Atmospheric Pressure Chemical Ionization and Exact-Mass Fourier Transform Mass Spectrometry

Craig S. Truebenbach, Hui Tong, Nelson Huang, Paul D. Schnier and Marshall M. Siegel

Wyeth Research, Chemical and Screening Sciences, Pearl River, NY 10965, USA

CONTENTS

High Throughput Analysis for Early Drug Discovery
Edited by James N. Kyranos

1 INTRODUCTION

Mass spectrometry (MS) using atmospheric pressure ionization methods is a powerful tool for obtaining molecular weight (MW) information of samples in a high throughput fashion and is ideally suited for characterizing the large number of samples produced using combinatorial chemistry synthesis technologies. A number of reports have appeared describing methods for high throughput analysis of single samples by electrospray ionization–mass spectrometry (ESI–MS) using flow-injection analysis (FIA) under low-resolution conditions. Very efficient methods were developed for automating the data acquisition, data processing, and data reporting processes.[1-9] In this report, an extension of the flow-injection technique is described for very high sample throughput utilizing a multi-sprayer ESI source (MUX) consisting of eight parallel sprayers for sequential ESI–MS data acquisition for the analysis of samples prepared in a 384-well plate format. Customized software for data processing and reporting of the MUX data is also illustrated. A modification of the MUX source for atmospheric pressure chemical ionization–mass spectrometry (APCI–MS) is described for the analysis of less polar samples produced using combinatorial chemistry synthesis technologies. The use of ESI Fourier transform ion cyclotron resonance mass spectrometry (FT-ICR MS or FTMS) for high-resolution exact-mass measurements of combinatorial chemistry samples, using automated high throughput data acquisition and data analysis modules, is described for confirming the elemental compositions of proposed structures with mass errors less than 0.5 ppm. Finally, a simple and efficient technique is described for cleaving compounds from combinatorial chemistry beads that is compatible with direct ESI mass spectral analysis.

2 HIGH THROUGHPUT MUX FLOW-INJECTION ESI AND APCI LOW-RESOLUTION MASS SPECTRAL ANALYSIS

Combinatorial chemistry parallel syntheses, followed by HPLC cleanup, can lead to a large number of fractions that may or may not contain the compound of interest. Conventionally, HPLC–ESI–MS of the fractions is used to determine the presence and purity of the compounds. However, even with the fastest HPLC gradients, this is a very time-consuming process. Since only a small percentage of the fractions contain the compounds of interest, most of the analysis time is unproductive. By flow-injection ESI–MS, the fractions that do not contain the compound can be quickly eliminated from further analysis. This is achieved by use of customized software that automatically analyzes and condenses the flow-injection data into an easy-to-read format, informing the chemist which fractions require further analysis by HPLC–ESI–MS. Using this screening protocol, only the limited numbers of fractions containing the components of interest are analyzed for purity by HPLC–ESI–MS in a high throughput and very efficient manner. Likewise, this same methodology was applied to less polar samples when the MUX ion source was modified for APCI–MS.

2.1   ESI Experimental Methods

The instrument used for the low-resolution FIA work is a Micromass LCT time-of-flight electrospray mass spectrometer equipped with an eight-way multi-sprayer source (MUX). Samples were injected into the mass spectrometer using a Gilson 215 autosampler equipped with a Gilson 889 injector with eight parallel needles and valves. The samples were submitted using 384-well plates, which is a convenient format for large numbers of samples.

Since hardware and software for the Gilson autosampler was designed for 96-well microtiter plates, it was necessary to modify the bed and plate setups in order to access samples in the 384-well plate. Well spacings are 9.0 mm apart for 96-well plates (an array of wells 8 columns × 12 rows), and 4.5 mm for 384-well plates (an array of wells 16 columns × 24 rows). Because of this difference, it was necessary to create two "half plates" with eight columns 9.0 mm apart to accommodate the needle spacing, and 24 rows spaced 4.5 mm apart. The two half plates were set up in the software so that the second plate was 4.5 mm to the left of the first plate thus effectively making a 384-well plate. This layout is shown in Fig. 4.1.

Two convenient methods were developed for preparing the samples in the 384-well plates and generating sample lists for the automated analysis of the mass spectral data. The most convenient method, if a robot was used to plate out the samples, was to program the robot to plate sequential samples into alternating wells. First, samples were loaded into "half plate 1," which corresponds to columns A, C, E,...,O of the 384-well plate, followed by "half plate 2," which corresponds to columns B, D, F,...,P of the 384-well plate. In this way, the first 192 samples were loaded into the first half plate and will be analyzed first. The sample list is created in a similar fashion, so that the sample information corresponded to the samples as they were arranged in each half plate. An alternate method for preparing the samples was to place them sequentially in the wells of the 384-well plate and to list them sequentially in a spreadsheet. The sample list was then reformatted to create "half plates 1 and 2." This was accomplished by creating in the spreadsheet a "sort column." In this column, alternating 1s and 2s were entered for alternating samples. The rows were then sorted based on the sort column. All samples with a 1 in the sort column corresponded to half plate 1 and will be in the first half of the list. All samples with a 2 in the sort column corresponded to half plate 2 and will be in the second half of the list.

Samples should be dissolved completely, leaving no residue, in a solvent, *i.e.* neither 100% organic, nor 100% aqueous since neither one electrosprays particularly well. Ideally, the solvent used should be identical to the HPLC carrier solvent so as to reduce the possibility of sample precipitation and eventual clogging of the instrument plumbing. High concentrations of additives such as dimethyl sulfoxide (DMSO) and trifluoroacetic acid (TFA) should be avoided since they ionize very efficiently in the positive and negative modes, respectively, thereby suppressing the analyte peaks. Likewise, analytes can also ion pair with TFA resulting in signal supression in the positive ion mode.[10–12]

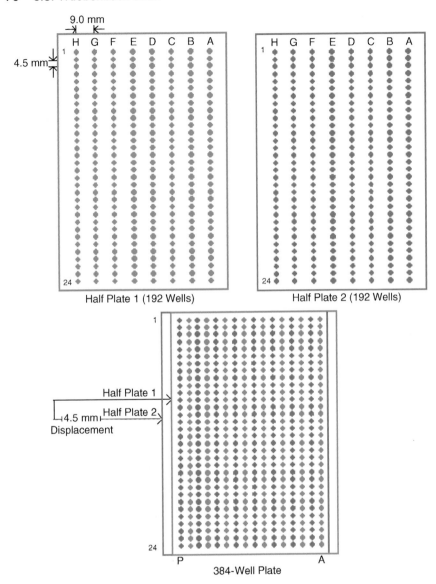

Figure 4.1   Schematic diagram showing two half plates (8 columns × 24 rows) each containing 192 wells used to create a 384-well plate (16 columns × 24 rows) by offsetting the two half plates by 4.5 mm. The dimensions for the well spacings are indicated.

Typical ESI MUX experimental conditions were as follows: capillary voltage 3000 V, cone voltage 20 V, MUX source and desolvation temperatures, 150 and 350°C, respectively, nitrogen nebulizer and desolvation gas flow rates, 650 and 700 l/h, respectively. 5 μl of the analyte was injected into the HPLC carrier solvent consisting of 1:1 water:acetonitrile (v/v) with 0.025% formic acid with a total flow rate of 800 μl/min, or 100 μl/min per sprayer. Once the sample was injected, data acquisition began immediately. Data were acquired for 30 s with 0.25 s per spectrum resulting in each channel being scanned once every 2 s. A total of 15 scans per channel were collected, the first five spectra being solvent background that was acquired before the analyte reached the mass spectrometer.

The overall data analysis process is illustrated in Fig. 4.2. Customized analysis software, referred to as the Analyzer, was used to automatically process the acquired data in batches of eight samples. This processing typically included combining the scans, subtracting the background, smoothing the data, subtracting the baseline and centroiding the data. The Analyzer software is very versatile and the data massaging modules can be added, deleted, rearranged or modified in any order.[1,2] The results of the Analyzer program were sent to an interpretation program which first subtracted the buffer/solvent background spectrum from the sample spectrum and then searched the data for peaks corresponding to molecular ion adducts of the compound of interest (in the positive ion mode: $[M + H]^{1+}$, $[M + Na]^{1+}$, $[2M + H]^{1+}$, etc., and in the negative ion mode: $[M - H]^{1-}$, $[2M - H]^{1-}$, $[M + Cl]^{1-}$, etc.). The signal-to-noise ratio and isotope fits were calculated for each of the detected molecular ion adducts. Automated data interpretation took about 30 min per plate. The results of the interpretation program were extracted by a third program that condensed the data and displayed the results in an Excel spreadsheet for review by the chemist.

2.2   ESI Results and Discussion

Using the above-described ESI-MS MUX technology, considerable time savings were achieved when analyzing samples produced by combinatorial chemistry parallel syntheses. The example described below and illustrated in Fig. 4.3 portrays this achievement. A chemist synthesized a library of 192 compounds, which were fractionated by preparative HPLC into 20 fractions per compound for a total of 3840 fractions. Initially, these samples were analyzed by HPLC-ESI-MS, however, most of the samples did not contain the compounds of interest, and the large stacks of data were very tedious to examine. This analytical approach was very inefficient and was terminated. The following approach was employed. For each compound, seven fractions with retention times most likely containing the compounds of interest were selected for further analysis, resulting in a total of 1344 samples. These samples were submitted for flow-injection ESI-MS MUX analysis in four 384-well plates. Overall acquisition and analysis time for all 1344 samples was 9 h. However, since the data from one plate could be processed while the data from another was being acquired, the total elapsed time for the analysis was about 6 h. The resulting data indicated that 150 of the fractions contained the compounds of interest, and those fractions were subsequently analyzed by HPLC-ESI-MS. When the time to analyze the 150 fractions by HPLC-ESI-MS is taken into account, the total time for analysis was reduced by about 20 times when compared to running all the samples by HPLC-ESI-MS. An illustration of the overall statistics is also described in Fig. 4.3.

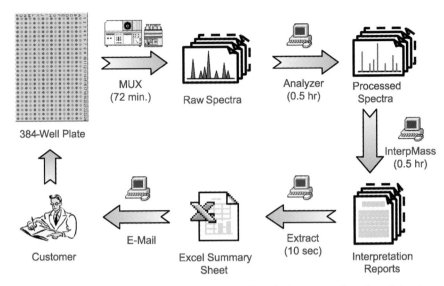

Figure 4.2   Overview of data flow for the overall analysis process. Samples originating from the chemist are automatically acquired using a multi-sprayer (MUX) system. The data are processed (combined, smoothed, baseline corrected, centroided, *etc.*) by the Analyzer program. The data are then processed by the InterpMass program which searches the data for peaks corresponding to the compound of interest (possible molecular ions and molecular ion adducts) and creates an interpretation report for each sample. Finally, the results are extracted and condensed into an Excel spreadsheet and e-mailed to the chemist. The time expended during each step in the assay of a 384-well plate by low-resolution MUX FIA ESI-MS is indicated below each arrow.

The automated flow-injection ESI-MS MUX technique had several other advantages in addition to the considerable reduction in analysis time. The acquisition and data processing were performed automatically with little operator interaction. Intense background peaks such as TFA were subtracted automatically, so that the weaker analyte peaks were easily identifiable. The resulting data were examined automatically for all likely molecular ion adducts which would have been very tedious and impractical to do manually. None of the spectra needed to be printed. The only printout was an Excel spreadsheet of the final results (Table 4.1). The analysis software significantly reduced the possibility of erroneous interpretations.

2.3   MUX Issues

Although use of a MUX multi-injector system has significant advantages in terms of throughput and analysis time, there are a number of issues that must be weighed when considering its overall practicality. Due to the complexity of the MUX system, great vigilance is required in its maintenance. There is a high probability for clogging and cross contamination since there are eight parallel channels. If one of

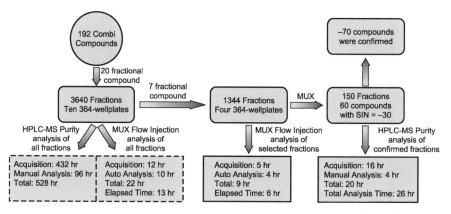

Figure 4.3   Overview of the statistics for the analysis of a large batch of combinatorial chemistry fractions described in the text. The initial batch of 3840 fractions, collected from 192 compounds, was reduced to 1344 fractions, of which 150 fractions had 70 compounds of interest. The time needed for each of the steps in the analyses are indicated within the rectangles at the bottom of the figure. The final analysis by HPLC-MS of 150 fractions took a total of 26 *vs* 528 h for the analysis of 3840 fractions, resulting in a 20-fold reduction in analysis time. (Rectangles with broken and solid borders correspond to extrapolated and experimental times, respectively.)

the channels partially or fully clogs, the solvent flow to that channel is diverted to the other seven channels with little change in system pressure. The only way to detect a problem is to monitor the mass spectral data, *viz.*, the signal intensity and retention time. Depending upon the severity of the clog, the signal could be completely lost and the retention time could increase considerably. If data acquisition is not monitored, there is the possibility of acquiring a large amount of bad data. Unless the data are observed closely over several injections, it is possible that a clogged channel could be mistaken for a bad sample. If unknowns are being analyzed, a problem can only be detected after observing several consecutive injections with no observed signal at a particular injector. With eight injection channels to monitor, noticing a trend in missing data can be a somewhat difficult task.

When a clog is detected, finding and resolving the problem can also be somewhat time-consuming. Because of the way in which the system is set up, there are five different locations in which a clog can occur. They are at the injector, the tubing to the injection valve, the injection valve, the tubing between the injection valve and the mass spectrometer, and the electrospray capillary. The best way to find the location of the clog is to swap tubing between the bad channel and a good channel starting at the mass spectrometer and working back toward the injector. For example, if channel one is clogged and all the rest are normal, swap the tubing from channel one and two. Run a standard and observe the mass spectral data as it evolves. If the error remains on channel one, the clog must be at the mass spectrometer source and most likely, the electrospray capillary needs to be changed.

Table 4.1 Customized Excel spreadsheet report for combinatorial chemistry samples (partial printout): data acquired from 384-well plates by low-resolution high throughput MUX flow-injection ESI-MS

File name	Well #	Notebook #	Elemental	MW (S/N total)[a]	M − H[b] (S/N)[c]	Isotope fit (%)	M + Others[b] (S/N)	Isotope fit (%)	M + Others (S/N)	Isotope fit (%)
p3_001	A.1	K2061-23C.3-1	$C_{24}H_{19}Cl_2N_1O_2$	423.079						
p3_002	C.1	K2061-23C.3-2	$C_{24}H_{19}Cl_2N_1O_2$	423.079 (4)			$[M+FA-H]^{1-}$ 468.1 (1)	0.0	$[M+TFA-H]^{1-}$ 536.2 (3)	66.5
p3_003	E.1	K2061-23C.3-3	$C_{24}H_{19}Cl_2N_1O_2$	423.079 (33)	$[M-H]^{1-}$ 422.3 (5)	44.4	$[M+FA-H]^{1-}$ 468.3 (2)	44.0	$[M+TFA-H]^{1-}$ 536.3 (26)	52.0
p3_004	G.1	K2061-23C.3-4	$C_{24}H_{19}Cl_2N_1O_2$	423.079 (703)	$[M-H]^{1-}$ 422.2 (101)	90.1	$[M+FA-H]^{1-}$ 468.2 (34)	73.5	$[M+TFA-H]^{1-}$ 536.2 (373)	95.0
p3_005	I.1	K2061-23C.3-5	$C_{24}H_{19}Cl_2N_1O_2$	423.079 (274)	$[M-H]^{1-}$ 422.2 (59)	91.3	$[M+FA-H]^{1-}$ 468.3 (19)	65.8	$[M+TFA-H]^{1-}$ 536.3 (196)	93.0
p3_006	K.1	K2061-23C.3-6	$C_{24}H_{19}Cl_2N_1O_2$	423.079 (36)	$[M-H]^{1-}$ 422.2 (8)	67.9	$[M+TFA-H]^{1-}$ 536.3 (28)	81.4		
p3_007	O.1	K2061-23C.3-7	$C_{24}H_{19}Cl_2N_1O_2$	423.079 (7)			$[M+TFA-H]^{1-}$ 536.3 (7)	69.5		
p3_008	A.2	K2061-23D.3-1	$C_{22}H_{18}Cl_1N_1O_2S_1$	395.074						
p3_009	C.2	K2061-23D.3-2	$C_{22}H_{18}Cl_1N_1O_2S_1$	395.074						
p3_010	C.2	K2061-23D.3-3	$C_{22}H_{18}Cl_1N_1O_2S_1$	395.074 (133)	$[M-H]^{1-}$ 394.2 (11)	72.4	$[M+FA-H]^{1-}$ 440.2 (4)	59.0	$[M+TFA-H]^{1-}$ 508.2 (84)	95.0
p3_011	E.2	K2061-23D.3-4	$C_{22}H_{18}Cl_1N_1O_2S_1$	395.074 (7)			$[M+TFA-H]^{1-}$ 508.2 (7)	66.1		
p3_012	G.2	K2061-23D.3-5	$C_{22}H_{18}Cl_1N_1O_2S_1$	395.074						
p3_013	I.2	K2061-23D.3-6	$C_{22}H_{18}Cl_1N_1O_2S_1$	395.074 (3)			$[M+TFA-H]^{1-}$ 508.3 (3)	51.9		
p3_014	K.2	K2061-23D.3-7	$C_{22}H_{18}Cl_1N_1O_2S_1$	395.074						
p3_015	M.2	K2061-23E.3-1	$C_{26}H_{21}N_1O_3$	395.152 (1)			$[M+TFA-H]^{1-}$ 508.3 (1)	39.8		
p3_016	O.2	K2061-23E.3-2	$C_{26}H_{21}N_1O_3$	395.152 (217)	$[M-H]^{1-}$ 394.3 (61)	0.0	$[M+FA-H]^{1-}$ 440.3 (7)	65.6	$[M+TFA-H]^{1-}$ 508.3 (149)	93.8
p3_017	A.3	K2061-23E.3-3	$C_{26}H_{21}N_1O_3$	395.152 (24)	$[M-H]^{1-}$ 394.3 (3)	0.0	$[M+FA-H]^{1-}$ 440.3 (1)	33.0	$[M+TFA-H]^{1-}$ 508.3 (20)	93.8
p3_018	C.3	K2061-23E.3-4	$C_{26}H_{21}N_1O_3$	395.152 (4)			$[M+TFA-H]^{1-}$ 508.3 (4)	50.6		
p3_019	E.3	K2061-23E.3-5	$C_{26}H_{21}N_1O_3$	395.152						
p3_020	G.3	K2061-23E.3-6	$C_{26}H_{21}N_1O_3$	395.152						
p3_021	I.3	K2061-23E.3-7	$C_{26}H_{21}N_1O_3$	395.152						

[a]Total S/N − total sum of signal-to-noise ratios (S/N) for all observed molecular ion adduct ions.

[b]M − H, M + Others − molecular ion adduct assignment and observed mass.

[c]S/N − signal-to-noise ratio of molecular ion adduct peak.

However, if the error moves to sprayer two, the clog has to be somewhere before the mass spectrometer. The next step is to move back towards the injector and swap the tubing at the next junction point. By troubleshooting the clog in this way, the location of the clog can be found and fixed.

Another issue with multi-injector systems is general mechanical wear-and-tear of the components. Because components are in multiples of eight, it is necessary to have a large number of spare parts available, and replacement of these parts is more frequent than with single injection systems. In general, the moving parts seem to wear out fairly regularly. These include the syringes, and the injection port seals. Also, depending on the type of needles and vial caps used, the injection needles can become clogged. If septum-piercing needles are used, over time small pieces of septa can accumulate in the injectors leading to a clog. Because of these potential problems, it is beneficial to set up a regular preventative maintenance schedule for replacing these parts. The timing of this maintenance will vary depending on the system usage, however, this does add significantly to the cost of operating the system.

All of the above issues add up to a large amount of maintenance when compared to single injection systems. If large numbers of samples are run on a regular basis, the increased throughput is certainly worth the added time and cost of the additional maintenance. However, the added maintenance must be taken into account when considering using a multi-injector system.

Another potential problem is the possibility of cross-talk. Since all samples are being sprayed simultaneously, if one particular sample is especially concentrated or ionizes particularly well, there is a possibility of cross-talk. This occurs when the signal from one sample carries over and appears on multiple channels. In extreme cases, the sample can appear on all channels and be strong enough to dwarf any analyte signal on that channel. This can lead to erroneous results, particularly if analytes with the same MW are being analyzed on adjacent sprayers. If one sample contains the analyte while the other does not, if there is cross-talk, it will appear that both samples contain the analyte. Even if the analytes are different, cross-talk can make it appear that there are other compounds in an otherwise pure sample.

Sample concentration can cause problems in other ways as well. The dynamic range of the micro-channel plate (MCP) detector is fairly narrow. The difference between the lower limit of detection for a particular sample and saturation of the detector is smaller than with other detection systems. This can lead to problems with quantitation, calibration, and to some extent, detection of weaker species. When an ion reaches the MCP detector and is counted, there is a lag time of about 5 ns before the next ion can be counted. At lower analyte concentrations/ion currents, this does not pose a problem. However, at high ion currents, a proportion of the ions generated are not counted leading to a shift to lower mass centroids and lower peak areas. This can be compensated for, to some extent, by using a "deadtime correction". However, it is best to avoid this problem of deadtime and detector saturation by adjusting sample concentration to lower values prior to analysis whenever possible.

2.4   APCI Experimental Methods

For less polar samples, atmospheric pressure chemical ionization (APCI) is the preferred ionization technique. Although the Micromass MUX system is designed for electrospray operation, it is possible to modify the MUX ESI source for APCI operation. APCI systems utilize a corona discharge needle, located in the flight path of vaporized analyte and solvent, for ionization of the sample. Vaporization of the analyte and solvent before reaching the corona discharge needle is achieved by using high temperatures in the source and high percentages of volatile organic solvents. The MUX source is very compact, leaving little room for additional components. However, corona discharge needles (World Precision Instruments, tungsten dissection needles PN 500134, 0.25 mm diameter), one per nozzle, can be attached to the gas manifold such that the tip of each APCI needle is directly in front of each nozzle (Fig. 4.4). For the most reliable APCI operation, the lengths of the electrospray nozzle capillaries were shortened from 8 to 4 mm. The additional space between the nozzle and MUX rotating baffle, in which the APCI needle resides, resulted in a significant reduction in arcing and allowed for more efficient nebulization, desolvation and chemical ionization. The voltage present on the gas manifold, 5000 V, was conducted to the APCI needle tips, creating a strong electric field resulting in corona discharges. To enhance solvent evaporation, the HPLC carrier solvent was enriched in organic composition to 95:5:0.025 acetonitrile:water:formic acid (v/v/v) and the flow rate was reduced to 200 µl/min total, equivalent to 25 µl/ min per sprayer, while maintaining the source and desolvation temperatures at 150 and 350°C, respectively. An external gas heater/controller (Analytica of Branford, Branford, CT, Model 101101) was added for heating the nebulizing gas and in turn the gas manifold to 300°C. The heated nebulizing gas aided in the vaporization of the analyte and solvent, increasing the ionization efficiency of the APCI process. The optimized nitrogen gas flow rates for nebulization and desolvation were 650 and 600 l/h, respectively. As a result of the lower carrier solvent flow rate used for the APCI system, longer acquisition times and lower throughput were achieved relative to the ESI only system. Nevertheless, even with the longer APCI acquisition times, by simultaneously analyzing 4 or 8 samples, the overall analysis time was decreased relative to a single APCI sprayer system.

2.5   APCI Results and Discussion

Preliminary APCI-MS results were obtained using the same MUX flow-injection ESI-MS system modified for APCI-MS, as described above. The APCI positive ionization mode was tested using the Agilent APCI tuning mix (PN 59987-60149) which produces four abundant ions in the positive ionization mode (m/z 121, 322, 622, and 1030). Three other compounds were used for testing the positive ionization APCI mode, *viz.*, 2[(phenylthio)methyl]-2-cyclopenten-1-one, thioxanthen-9-one,

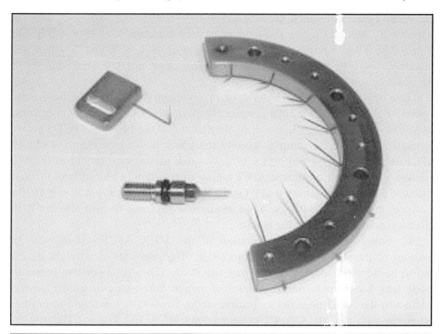

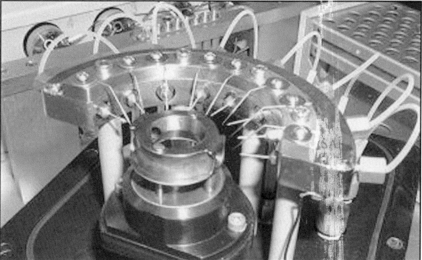

Figure 4.4   MUX source components and modified MUX system for APCI. The upper panel shows a single ESI nozzle, which is modified for APCI by shortening the capillary tip from 8 to 4 mm, a single APCI needle and a cradle holding eight APCI needles. The lower panel illustrates the cradle mounted on the gas manifold to which the nozzles are attached. The eight APCI needles are positioned directly in front of all eight modified ESI nozzles providing a corona discharge for each sample channel. The circular device is the MUX rotating baffle with two sampling orifices. As the baffle rotates, an orifice aligns before each APCI needle and nozzle transmitting the APCI plasma to the sampling cone of the mass spectrometer while the baffle blocks transmission from all other APCI needles and nozzles.

and methyl stearate, which produce $[M + H]^{1+}$ ions at m/z 205, 213, and 299, respectively. Methyl stearate also produced a very abundant $[M + H + ACN]^{1+}$ ion at m/z 340. These samples were analyzed on nozzles 1, 3, 5, and 7, which were shortened and fitted with APCI needles. The other nozzles were kept in the electrospray configuration.

Figure 4.5 illustrates the mass spectra obtained simultaneously by positive ionization APCI using the four samples listed above. Although these compounds also produced spectra by electrospray, the spectra shown here are APCI. This was demonstrated when the samples were re-run under the same conditions with the APCI needles removed. No peaks, or very weak peaks, were produced from the samples except for the Agilent APCI tuning mix, which produced two distinctly different peak distributions by APCI and ESI with an abundant ion at m/z 1030 in the spectrum acquired with the APCI needle installed, and its near absence in the spectrum acquired without the needle.

Two issues related to the operation of the MUX APCI system were the position of the APCI needles and cross-talk. The positions of all eight needles had to be adjusted for optimum signal sensitivity. The APCI positive ionization mode was found to be more tolerant of minor differences in needle position relative to the APCI negative ionization mode. Cross-talk was found to be more extensive with the MUX APCI system *vs* the MUX ESI system. The simplest way found to compensate for cross-talk was to use slightly different lengths of PEEK tubing for each of the eight sprayers, so that the elution times of the samples were staggered. When processing the data, spectra were summed in the region where the analyte was expected to elute, and subtracted from all other regions. In this way, most of the cross-talk was eliminated from the processed APCI data.

A number of enhancements can be envisioned for improving the MUX source for APCI. A single centralized needle would improve reliability and a wider baffle would reduce the residence time of stray ions around the inlet cone, thereby reducing cross-talk. As new instrumentation is developed, these proposed improvements would most likely be introduced. Nevertheless, the MUX APCI system, as described above using staggered elution times, is functional for FIA of less polar small molecules that are not amenable to ESI analysis. A viable alternative mode to MUX APCI referred to as "solvent stream selection" was recently demonstrated by Covey et al.[13] whereby solvent streams are individually turned on sequentially for analysis by APCI or ESI.

3   HIGH THROUGHPUT HIGH-RESOLUTION EXACT-MASS FIA ESI-FTMS

Large numbers of samples are produced using parallel synthesis combinatorial chemistry procedures. These materials are either produced in-house or obtained from outside combinatorial library vendors. Generally, these samples, after purification, are prepared in limited quantities and deposited in well-tray formats,

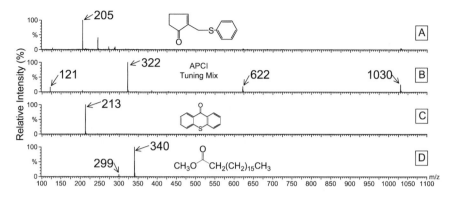

Figure 4.5 Simultaneously acquired APCI spectra using a modified Micromass MUX electrospray source. The spectra shown are of the following samples: (A) 2[(phenylthio)-methyl]-2-cyclopenten-1-one with an $[M + H]^{1+}$ ion at m/z 205, (B) Agilent APCI tuning mix with $[M + H]^{1+}$ ions at m/z 121.1, 322, 622, and 1030, (C) thioxanthen-9-one with an $[M + H]^{1+}$ ion at m/z 213, and (D) methyl stearate with an $[M + H]^{1+}$ ion at m/z 299 and with an $[M + H + ACN]^{1+}$ ion at m/z 340.

often of low volumes at a concentration of 30 mM in pure DMSO. The chemical structures have to be characterized and validated for new records of invention and eventually for patent applications. Often these samples produce abundant molecular ions using ESI–MS in either the positive or negative ionization modes. Under high-resolution exact-mass conditions using flow-injection Fourier transform ion cyclotron resonance mass spectrometry (FT-ICR MS or FTMS), reliable masses for the molecules can be measured to four decimal places permitting the determination of the unique elemental formula for the compound, thereby validating the proposed synthesis of the product molecule. Due to the high sensitivity of the electrospray method, very small quantities of the materials are used leaving enough material for biological screens. In addition, since the FTMS measurements and interpretations are fully automated, the high throughput assays are completed in a very reliable manner. For these reasons, FTMS is ideally suited for compound validation and for records of invention and to our knowledge is the best single analytical method for achieving these goals. Recently, a review dealing with FTMS high-resolution applications in combinatorial chemistry has appeared.[20]

3.1 High-Resolution Experimental Methods

The FT mass spectrometer used was a Bruker Daltonics APEX II equipped with a passively shielded 9.4 T superconducting magnet, an external Analytica ESI source, an Agilent Model 1100 HPLC system and a Silicon Graphics O2 Workstation data system. The instrument was operated as previously described.[21] The samples were received at a concentration of 30 mM in DMSO and were diluted to 0.3 mM with acetonitrile for ESI-FTMS analysis. One μl was injected with a carrier solvent flow rate of 100 μl/min of 1:1 acetonitrile:water (v/v) with 0.025% formic acid. Samples were run every 3 min. Typically 512K data points were acquired. For low MW combinatorial chemistry samples of pharmaceutical interest, resolving powers of 50 000 (FWHM) and absolute mass errors less than 0.5 ppm were routinely achieved with external calibration.

Figure 4.6 is a flow diagram illustrating the two independent automation modules used for data acquisition (left side) and data processing (right side) of ESI FT mass spectra for the combinatorial chemistry samples.[21] Data acquisition was initiated by a contact closure signal from the Agilent autosampler which triggers the Bruker data system to acquire spectra for a fixed number of acquisitions of about 1 s each, after an initial fixed delay. This process was repeated for all the samples every 3 min. After completing data acquisition, the data processing module was loaded with a spreadsheet containing the sample information (request number, structure number, proposed elemental formula and proposed exact MW, *etc.*) which was compiled manually or from the corporate database. The acquired time-domain data were processed and transformed into a peak-picked mass spectrum. Using the proposed elemental formula for each sample, all the expected molecular ion adduct masses were computed and correlated with the observed exact-masses. For each sample, a report was generated that lists all the correlated ions within a 3 ppm window of the predicted masses and the spectrum was automatically printed out, if desired.

3.2   High-Resolution Results and Discussion

Table 4.2 illustrates a partial printout of the ESI-FTMS high-resolution exact-mass report customized for combinatorial chemistry samples. This report was designed so that each line of the spreadsheet gives sufficient information for each sample so that it is not necessary to view each spectrum and exact-mass report to verify the consistency of the proposed composition with the observed mass spectral results. The first three columns contain the sample descriptors, *viz.*, vial number, request number and chemical library structure number. This is followed by the predicted elemental formula, most significant molecular ion adduct observed with its predicted and observed masses, as well as the mass error between the observed and predicted masses in millimass and ppm units (note again that the mass errors are

---

Figure 4.6   Schematic diagram of the hardware and software used for automated data acquisition, processing and e-mailing of high-resolution exact-mass flow-injection ESI-FTMS data, using a Bruker 9.4 T Apex II FTICR MS equipped with a Agilent 1100 HPLC system. The upper left side of the figure indicates the automated data acquisition process that is initiated by a contact closure from the Agilent 1100 Autosampler. The upper right side of the figure illustrates the initial steps taken for automated data processing, *viz.*, the creation of a sample list which is then associated with the processed and interpreted FTMS data, resulting in exact-mass reports.

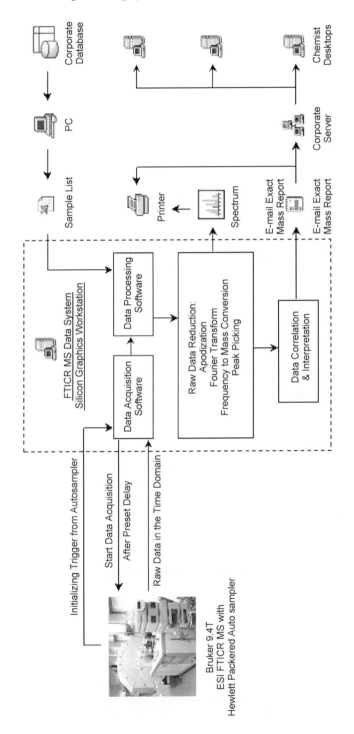

consistently less than 0.5 ppm using this ESI-FTMS system). In addition, the signal-to-noise ratio of the most significant molecular ion adduct and the mass spectral purity of the sample are tabulated. Other less significant molecular ion adducts may have been observed but are not reported in this summary table. Such information is available in the complete exact-mass reports. The mass spectral purity is defined as the sums of all the ion currents for all the interpreted ions as well as all their corresponding isotopes within 3 ppm divided by the total ion current for the sample. The proposed structures are considered to be validated for the combinatorial chemistry used when the signal-to-noise ratio for the most significant molecular ion and the mass spectral purity are high. However, these exact-mass spectral results only verify that the observed structures are consistent with the proposed elementals but do not verify that the proposed isomeric forms of the molecules were generated.

4   SAMPLE PREPARATION METHODS FOR ESI-MS ANALYSIS OF DRUGS BOUND TO RESIN BEADS

When synthesizing samples on beads, a method for quickly and conveniently monitoring the progress of the reaction is very useful. One of the best tools for tracking a reaction in solution is by sampling the reaction solution and analyzing it by ESI–MS throughout the course of the reaction. This approach is somewhat more difficult in the case of synthesis on beads. The products or chemical intermediates have to be cleaved from the beads prior to analysis without any degradation, a procedure normally applied as the last step in the synthesis. Ideally, small amounts of the beads would be needed for the analysis, reducing the impact on percent yield. In order to use the smallest number of beads, an efficient method of extraction is necessary. What follows are two techniques which use a minimal amount of beads, and by having one or no sample transfer steps, sample loss is kept to a minimum. The ESI mass spectral approach of analyzing the cleaved compound is simpler, more sensitive and rapid than the solid state *in situ* approach of analyzing resin bound compounds by solid-state NMR[22–25] and IR[26–28] spectroscopies. The electrospray mass spectra are also easier to interpret than the NMR and IR data.

4.1   Sample Preparation Experimental Methods

The first technique has one sample transfer step, and uses standard autosampler vials. Approximately, 100 beads were placed into an autosampler vial. Ten microliters of 10% TFA in CH_2Cl_2 (v/v) was added. The vial was loosely capped and placed in a hood for 5 min. This was followed by the addition of 30 μl of formic acid and 60 μl of acetonitrile. The supernatant was transferred to a new autosampler vial using an ultra-micro-Eppendorf pipette tip (Geloader tip) piggybacked with a 100 μl pipette. The Geloader tip serves as a particulate filter. This new sample could then be analyzed by any standard electrospray mass spectrometer.

Table 4.2 Customized Excel spreadsheet report for combinatorial chemistry samples (partial printout): high-resolution exact-mass data acquired automatically by flow-injection ESI-FTMS

Vial #	Request #	Structure #	Elemental formula	Adduct	Predicted mass (Da)	Experimental mass (Da)	Error (mmu)[a]	Error (ppm)[b]	S/N[c]	Purity[d] (%)
1	281989	WAC-65515186	C.32.H.28.Cl.1.N.1.O.4	$[M-H]^{1-}$	524.16341	524.16347	0.06	0.11	160	63.78
2	281990	WAC-65522169	C.25.H.19.N.1.O.4.S.1	$[M-H]^{1-}$	428.09620	428.09630	0.10	0.23	128	88.44
3	281991	WAC-65522170	C.29.H.21.N.1.O.5	$[M-H]^{1-}$	462.13469	462.13477	0.08	0.17	125	87.05
4	281992	WAC-65522171	C.34.H.25.Cl.2.N.1.O.4	$[M-H]^{1-}$	580.10878	580.10875	-0.03	-0.05	118	86.79
5	281993	WAC-65522172	C.32.H.25.N.1.O.4	$[M-H]^{1-}$	486.17108	486.17110	0.02	0.04	130	87.32
6	281994	WAC-65522173	C.34.H.25.N.1.O.4	$[M-H]^{1-}$	510.17108	510.17098	-0.10	-0.20	130	83.34
7	281995	WAC-65522174	C.30.H.25.N.2.O.4	$[M-H]^{1-}$	473.15068	473.15071	0.03	0.06	68	96.43
8	281996	WAC-65522176	C.34.H.25.N.1.O.4	$[M-H]^{1-}$	510.17108	NOT OBSERVED			0	0.00
9	281997	WAC-65522177	C.31.H.22.F.1.N.1.O.4	$[M-H]^{1-}$	490.14601	490.14609	0.08	0.16	231	66.87
10	281998	WAC-65522178	C.36.H.31.N.1.O.6	$[M-H]^{1-}$	572.20786	572.20805	0.19	0.33	595	86.91
11	281999	WAC-65522179	C.34.H.25.N.1.O.4	$[M-H]^{1-}$	510.17108	510.17104	-0.04	-0.08	193	80.46
12	282000	WAC-65522180	C.35.H.27.N.1.O.5	$[M-H]^{1-}$	540.18164	540.18178	0.14	0.26	603	79.57
13	282001	WAC-65522181	C.34.H.27.N.1.O.5	$[M-H]^{1-}$	528.18164	528.18170	0.06	0.11	366	84.73
14	282002	WAC-65522182	C.35.H.37.N.1.O.4	$[M-H]^{1-}$	534.26498	534.26490	-0.08	-0.15	176	85.81
15	282003	WAC-65522183	C.31.H.22.Br.1.N.1.O.4	$[M-H]^{1-}$	550.06594	550.06596	0.02	0.04	138	79.61
						Average error	0.04	0.08		
						Average absolute error	0.07	0.14		
						Standard deviation	0.08	0.15		

mmu, millimass units; ppm, parts per million.

[a] Error (mmu) = experimental mass (mmu) − predicted mass (mmu).

[b] Error (ppm) = error (mmu) × 10³/predicted mass (Da).

[c] S/N – signal-to-noise ratio of molecular ion adduct peak.

[d] Purity – mass spectral purity (see text).

The second technique required no sample transfers, however, a special vial was required. The Whatman Mini-UniPrep Syringeless Filter system consists of an outer cup and an inner plug containing a filter on one end and a snap cap with septum on the other end. When assembled, the system becomes an autosampler vial. The disassembled Mini-UniPrep system was placed in acetonitrile and sonicated for 3 min. This process was repeated with fresh solvent (this removes PEGs from the plastic). Approximately 100 beads were placed into the outer cup of the Mini-UniPrep assembly followed by 10 μl of 10% TFA in CH_2Cl_2, (v/v) and allowed to stand for 5 min in a hood with the cap loosely attached (tilting of the outer cup enabled most of the beads to be wetted with the small volume of solvent). After 5 min has passed, 30 μl of formic acid and 150 μl of acetonitrile were added. The system was then assembled by inserting the inner plug into the outer cup by pressing the two pieces together. When inserting the inner plug, the solution was filtered, leaving the beads in the outer cup and the filtrate in the inner cup. It was important that two clicks be heard when assembling the vial. The inner plug MUST touch the bottom of the outer cup. The two parts were similar to a standard autosampler vial and could be used in any standard electrospray mass spectrometer.

4.2   Sample Preparation Results and Discussion

Two experiments were performed to evaluate and optimize a micro-extraction procedure using approximately 100 beads. Ten micro-liters of different concentrations of $TFA:CH_2Cl_2$ solutions were used to extract known bound moieties from the beads. After 10 min, each sample was diluted with 30 μl of formic acid and 60 μl acetonitrile. Maximum extraction was observed with $\geq 10\%$ TFA in CH_2Cl_2 (v/v). A time course study was then undertaken with 10% TFA in CH_2Cl_2. It was found that after 5 min maximum extraction was achieved. The extraction procedure was very simple since it requires either no or one sample transfer step and was very rapid. In all cases, the ESI mass spectra in the positive ion mode were of high quality with abundant signal. Large numbers of samples can be analyzed rapidly and routinely in the open access MS mode.

The procedure for cleaving drugs from beads, described above, was developed based upon the following chemical and mass spectral rationalizations. High concentrations of TFA in CH_2Cl_2 were used in the first step since a high concentration of TFA was required to cleave the compounds from the beads while the CH_2Cl_2 served as the extraction solvent for the compound. At these high TFA concentrations, analyte ion production was suppressed in the ESI positive ionization mode by ion pairing and in the ESI negative ionization mode by the high abundance of TFA anions.[10-12] For these reasons, a second step was introduced whereby the TFA concentration was drastically diluted with strong organic acid (formic acid) and a polar organic solvent (acetonitrile), enabling the production of abundant analyte ions in the ESI positive and negative ionization modes.

To test this methodology, four samples were prepared following the procedures described above using autosampler vials with conical inserts and Mini-UniPrep Syringeless Filter vials. In both cases, the mass spectral results were identical with those obtained using a time-consuming extraction procedure used by combinatorial chemists for preparing macro-quantities of resin-free final products.

5   CONCLUSIONS

By using a combination of high throughput hardware and software tools, it was possible to significantly reduce analysis times for large numbers of combinatorial chemistry samples. A mass spectrometer equipped with a parallel multi-sprayer interface with ESI and APCI capabilities allowed for very rapid analysis of large numbers of samples. When this was followed by automated data reduction and interpretation, results were obtained routinely in a very efficient manner. The screening of fractions by parallel flow-injection MS allowed the chemist to focus attention principally on the fractions that contained the product of interest, making the overall analysis process much more efficient, reducing the overall analysis time by a factor of about 20 times for the examples cited above.

Alternately, commercial systems are available which use a mass spectrometer as a detector to control which LC fractions are collected.[14–19] These systems can collect fractions in two modes. When the mass of the analyte is known, the system will only collect fractions containing the mass of interest. If the mass is not known, it will collect fractions for all detected UV peaks and automatically acquire mass spectra for each of the fractions. The primary drawback of such a system is the limited throughput and inefficient use of instrumentation. Because the mass spectrometer is dedicated to the analysis of the LC eluent, a large portion of instrumentation time is utilized in acquiring a small amount of data. In order to increase throughput with this type of system, it is necessary to buy more systems including the mass spectrometer, a relatively expensive proposition. By dividing the analysis process into the individual steps of chromatography, fractionation and MS, more options in instrumentation usage and method development are possible. Since the mass spectrometric analysis is significantly faster than the LC separation, increasing the number of relatively inexpensive preparative LC systems can increase throughput. In this way, fractions from many preparative LC systems can be analyzed using a single mass spectrometer. Although the analysis is not completed using one integrated set of instruments, each step can be optimized increasing efficiency and reliability even further. The main challenge with the eight-way injector system was to keep it working in a highly reliable manner. Due to the complexity of the plumbing, regular preventative maintenance was necessary to minimize leaking and clogging of the system.

In addition to the MUX low-resolution FIA–MS system for obtaining molecular ion information, an automated high throughput high-resolution exact-mass ESI-FTMS system has been developed for confirming the elemental formulas for large numbers of compounds prepared in parallel synthesis combinatorial chemistry programs. The system utilizes a minimum amount of sample and the observed data are valuable for validating the proposed elemental compositions for records of invention and patent purposes. Customized software was developed to display in spreadsheet format a one-line report, for each sample, listing the most significant molecular ion observed and its measured signal-to-noise ratio, the mass error between the observed and predicted mass and the mass spectral purity of the sample. Typically, mass errors of less than 0.5 ppm are observed. Finally, an efficient technique was described for cleaving compounds from combinatorial chemistry beads that is compatible with direct ESI mass spectral analysis.

ACKNOWLEDGEMENTS

The authors greatly appreciate the many helpful comments and insights offered by Scott Kincaid, Marc Papi, Gene Trybulski, Kevin Olsen, Keiko Tabei, William DeMaio and Guy Carter in preparing the samples and developing the methods used in this work.

REFERENCES

1. H. Tong, D. Bell, K. Tabei, and M.M. Siegel, *J. Am. Soc. Mass Spectrom.*, **10**(11), 1174–1187 (1999).
2. H. Tong, K. Tabei, F. Moy, R. Powers, M.M. Siegel, *Proceedings of the American Society for Mass Spectrometry and Allied Topics*, 48th ASMS Conference, June 11–16, 2000, 935–936 (WPH320) (2000).
3. G. Hegy, E. Goerlach, R. Richmond, and F. Bitsch, *Rapid Commun. Mass Spectrom.*, **10**(15), 1894–1900 (1996).
4. E. Goerlach, R. Richmond, and I. Lewis, *Anal. Chem.*, **70**(15), 3227–3234 (1998).
5. E. Goerlach and R. Richmond, *Anal. Chem.*, **71**(24), 5557–5562 (1999).
6. R. Richmond, E. Gorlach, and J.-M. Seifert, *J. Chromatogr.*, **835**(1 + 2), 29–39 (1999).
7. R. Richmond and E. Gorlach, *Anal. Chim. Acta*, **390**(1–3), 175–183 (1999).
8. R. Richmond and E. Gorlach, *Anal. Chim. Acta*, **394**(1), 33–42 (1999).
9. N. Yates, D. Wislocki, A. Roberts, S. Berk, T. Klatt, D.-M. Shen, C. Willoughby, K. Rosauer, K. Chapman, and P. Griffin, *Anal. Chem.*, **73**(13), 2941–2951 (2001).
10. F.E. Kuhlmann, A. Apffel, S.M. Fischer, G. Goldberg, and P.C. Goodley, *J. Am. Soc. Mass Spectrom.*, **6**(12), 1221–1225 (1995).
11. A. Apffel, S. Fischer, G. Goldberg, P.C. Goodley, and F.E. Kuhlmann, *J. Chromatogr. A*, **712**(1), 177–190 (1995).
12. J. Eshraghi and S.K. Chowdhury, *Anal. Chem.*, **65**(23), 3528–3533 (1993).
13. T. Covey, A. Weiss, R. Jong, *Symposium on Chemical and Pharmaceutical Structure Analysis (CPSA)*, Princeton, NJ, October 9–11 (2001)
14. L. Zeng, L. Burton, K. Yung, B. Shushan, and D.B. Kassel, *J. Chromatogr. A*, **794**(1 + 2), 3–13 (1998).
15. L. Zeng and D.B. Kassel, *Anal. Chem.*, **70**(20), 4380–4388 (1998).
16. D.M. Drexler and P.R. Tiller, *Rapid Commun. Mass Spectrom.*, **12**(13), 895–900 (1998).
17. J.P. Kiplinger, R.O. Cole, S. Robinson, E.J. Roskamp, R.S. Ware, H.J. O'connell, A. Brailsford, and J. Batt, *Rapid Commun. Mass Spectrom.*, **12**(10), 658–664 (1998).

18. G.J. Dear, R.S. Plumb, B.C. Sweatman, I.M. Ismail, and J. Ayrton, *Rapid Commun. Mass Spectrom.*, **13**(10), 886–894 (1999).
19. R. Maiefski, D. Wendell, W.C. Ripka, and J.D. Krakover, U.S. Pat. Appl. Publ., 43 pp., Division of U.S. Ser. No. 430,194 (2001).
20. D.G. Schmid, P. Grosche, H. Bandel, and G. Jung, *Biotechnol. Bioeng.*, **71**(2), 149–161 (2001).
21. N. Huang, M.M. Siegel, G.H. Kruppa, and F.H. Laukien, *J. Am. Mass Spectrom.*, **10**(11), 1166–1173 (1999).
22. G. Lippens, R. Warrass, J.-M. Wieruszeski, P. Rousselot-Pailley, and G. Chessari, *Comb. Chem. High Throughput Screening*, **4**(4), 333–351 (2001).
23. M.J. Shapiro, J. Chin, R.E. Marti, and M.A. Jarosinski, *Tetrahedron Lett.*, **38**(8), 1333–1336 (1997).
24. P.A. Keifer, L. Baltusis, D.M. Rice, A.A. Tymiak, and J.N. Shoolery, *J. Magn. Reson., Ser. A*, **119**(1), 65–75 (1996).
25. R.S. Garigipati, B. Adams, J.L. Adams, and S.K. Sarkar, *J. Org. Chem.*, **61**(8), 2911–2914 (1996).
26. H. Bandel, W. Haap, and G. Jung, *Comb. Chem.*, 479–498 (1999).
27. B. Yan, H.-U. Gremlich, S. Moss, G.M. Coppola, Q. Sun, and L. Liu, *J. Comb. Chem.*, **1**(1), 46–54 (1999).
28. B. Yan, J.B. Fell, and G. Kumaravel, *J. Org. Chem.*, **61**(21), 7467–7472 (1996).

5

Purity and Quantity Determination of Parallel Synthesis Compound Libraries

Jeannine Delaney

Pfizer Discovery Technology Center, 620 Memorial Drive, Cambridge, MA 02139, USA

Lesline V. Julien and James N. Kyranos

ArQule, Inc., 19, Presidential Way, Woburn, MA 01801, USA

Christine Salvatore

SRI International, 4111 Broad Street, San Luis Obispo, CA 93401-7903, USA

CONTENTS

High Throughput Analysis for Early Drug Discovery
Edited by James N. Kyranos

1   INTRODUCTION

Combinatorial chemistry encompasses a wide variety of synthetic approaches ultimately designed to create diverse libraries of compounds with the promise that some molecules will successfully target diseases and lead to new medicines.[1,2] Two of these synthesis techniques are split-and-pool and parallel synthesis. The main difference between the two is that split-and-pool approaches attempt to maximize molecular diversity within a single reaction vessel while parallel synthesis strategies aim to create a single product per well. Both approaches have advantages and disadvantages, which must be weighed with respect to the desired outcome of a particular drug discovery process. For example, split-and-pool methods maximize the diversity of molecules that can be initially screened against a target, however, wells displaying activity must then be de-convoluted and the active molecule isolated from the complex matrix of the other compounds present in the reaction mixtures. Although parallel synthesis is perhaps the most labor intensive of the combinatorial processes with respect to compound preparation and analysis, there is a significant advantage since it can be used to produce compounds that can be screened for initial hits and used directly in late-stage studies.[2] Regardless of the production method, ultimately compounds must be positively identified and of high quality and have sufficient purity and quantity to not only identify hits against disease targets but to support late-stage drug discovery studies including toxicology and pharmacokinetics.[2]

One of the first companies to develop and successfully implement automated high throughput parallel synthesis of several hundred thousand single compounds per year in a spatially addressable format was ArQule, Inc.[3] The characterization of compounds produced in such an automated high throughput process places significant demands on laboratory workflow, chromatographic methodology, instrument capacity, and data capture, processing and storage. To address the requirements of characterization in an efficient and cost-effective manner, a defined analytical strategy must be developed and employed.

2   LIBRARY DEVELOPMENT AND PRODUCTION

To provide a basis for understanding the size and scope of parallel synthesis characterization, the processes of library development and production are outlined. The first step of library development is to optimize probe chemistry and establish general reaction conditions in a manner similar to traditional medicinal chemistry approaches. The probe chemistry is worked up from a few experimental vials to a reaction set that examines a cross section of the desired products.

Historically, parallel synthesis has been performed in a combinatorial cross pattern: X by Y by Z where X, Y, and Z designate grid coordinates. For example, X may be a reagent list of amines that will be crossed with Y, a reagent list of aldehydes, for each molecular scaffold or template, Z.

Beyond the initial probe chemistry to establish a suitable list of compounds for library production, a "reagent qualification" is performed. All the reagents designated for one axis are crossed with reagent(s) previously identified as suitable for the chemistry to qualify them for production use. In the previous example, an amine qualification would be achieved by crossing a selected aldehyde, or a small well-characterized subset of aldehydes, with all potential amines. Conversely, crossing all desired aldehydes with a small subset of designated amines would perform an aldehyde qualification. In both examples, the same molecular scaffold(s) would be used in order to verify the overall reactivity and yields.

An important look at the general applicability of the reaction conditions to a diverse set of compounds is achieved by creating a "test plate." This is a process of selecting X, Y and Z components for a diverse subset of products that includes the maximum variation in chemical properties. Ideally the test plate should be created in an automated production environment using robotic workstations,[4] so that the results are an indication of the performance of the larger reagent set and adjustments to the protocol can be made prior to full-scale library synthesis.

Finally, a parallel synthesis library is the full multi-dimensional combinatorial matrix of compounds from a single medicinal chemistry idea projected onto a 2D plane of reaction tubes. Each library is further divided into subsets called arrays, which represent a group of samples always moving through the synthesis process as a distinct entity. In a more traditional manufacturing process, an array can be thought of as a batch of compounds. The array is divided further by run and block where run designates the time of production, and block refers to the rack of vials produced during a particular run. A block may consist of a 4 by 6 reagent cross (24-well), or an 8 by 12 cross (96-well). Although there is currently a trend to move towards higher well density as high throughput technology develops, particularly as it relates to high throughput screening of biological targets, the 24-well rack is still the best approach for synthesis. The available volume in one of the tubes used in the 24-well block is sufficient for efficient development of reaction conditions to synthesize approximately 20 mg of each compound, while the physical dimensions of the tubes optimize the number of compounds that can run together as a batch. Several runs of a library may be beneficial to gain production experience with a particular type of chemistry, and to expand the lists of reagents and templates while keeping the over-all scale of the process manageable. Library compounds are dried following production and then reconstituted to a predetermined concentration in DMSO, which generally dissolves most compounds and is compatible with most biological assays. The QC dilution scheme is dependent on the target concentration of the products and is optimized for the analytical instrumentation.

3 ANALYTICAL PROCESS

A central analytical group supports characterization and purification during the different stages of chemistry development and library production at ArQule. Having one analytical group support characterization and purification regardless of the chemistry/production stage provides a way to efficiently utilize resources. Library samples are prepared using a subset of reagents; thus, there is inherent homology in the molecules and the use of standardized analytical methods is appropriate. Because standard analytical methods are used which can be applied to development vials and production plates alike, instruments can be loaded to capacity. In addition, characterization samples are dispensed into shallow well polypropylene plates that hold micro-liter quantities (150 μl), which decrease solvent consumption and spatial requirements for analysis and storage.

As previously indicated, the general high throughput synthesis process may be generally defined as early development, reagent qualification, test plate and production and for each stage of the process there is a corresponding analytical component to ensure the overall quality of the products at that stage. For example, during a project's early development, while the reaction chemistry is being probed, NMR is used quite extensively and the correlation of this technique with general HPLC/MS analysis is validated. General high throughput HPLC/MS methods can then be applied as the principle analytical approach and the results used to rapidly optimize product synthesis. Mass spectral confirmation is the initial pass/fail criteria applied and indicates the presence of the desired products. Purity and quantity are not critically evaluated and the choice of the HPLC/MS method is not yet finalized.

During the next stages of library development, reagent qualification and test plate, analytical results are used to help make decisions about the specific reagents that will be used and the likelihood of production success with the reaction scheme, which may continue to be optimized at this point. The appropriate HPLC/MS methods are determined and modified if necessary, as these will be applied throughout the remainder of the process. The data are critically evaluated to ensure that products can be separated from by-products and impurities and that mass spectral response is adequate. Complete characterization, which for these purposes may be defined as the determination of purity, quantity and positive mass confirmation, is performed on reagent qualifications and test plates. Moreover, samples derived from these stages undergo HPLC/MS purification prior to being evaluated for purity and quantity. Primary purity assessment is performed using evaporative light scattering detection (ELSD) and UV at 214 nm as a secondary assessment technique. Similarly, primary quantity determination is performed using gravimetric analysis and ELSD response is used as a confirmatory method.

Individual production samples may undergo a series of analyses, which include pre-purification, post-purification, post-reformat and sample sub-setting confir-mation before being ready to be placed in the repository or screening assay. Each analysis round has its own purpose and sampling requirements. A statistical sampling of products prior to purification validates the chemistry and the automation used in its production. For this reason, an initial round of QC is performed on 25% of the samples prior to library purification. Figure 5.1 shows a software representation screen of a pre-purification sampling pattern obtained from our in-house synthesis tracking system. The white wells indicate samples that will be pulled for analysis. The diagonal pattern is a cross-section of the production

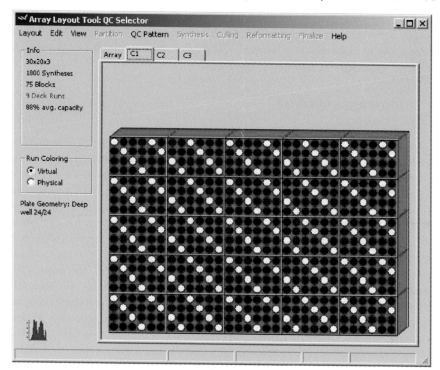

Figure 5.1 Representative 25% pre-purification sampling pattern of a library run. The white wells indicate the samples designated for QC analyses.

run and ensures that samples are taken from every column of the matrix for analysis, in order to understand the performance of each reagent used. While these samples are impure, it is possible to estimate their quantity using ELSD detection. In addition, a comparison of pre- and post-purification results can be used to assess the effectiveness of the overall purification process.

Purification is the most costly step in terms of time and solvent consumption; therefore, if a low percentage of products pass pre-purification criteria, corrective action or alternate purification strategies may be appropriate. If the percentage of compounds passing pre-purification criteria is high, a 100% sample purification strategy is adopted.

Post-purification analyses confirm product quality by providing a measure of purity and quantity and the criteria for acceptable purity and quantity are set based on the needs of the discovery value chain that the products are intended for. For instance, compounds intended for initial target screening may have a different purity and quantity criteria than compounds that will be used for structural activity relationship expansion in early lead optimization. Post-purification samples that fail the final product criteria are culled or physically removed from the batch.

Post-reformat QC is used to check that the sample compression after culling was performed correctly. A small number of samples are analyzed from each block and the resulting data quickly identifies if reformatting errors had occurred due to human error or instrument malfunction. Subsequent sample handling of the compounds to other plates or storage vessels may have additional analysis steps designed to assure the validity of the process.

4 PURITY ASSESSMENT

There are two common approaches for obtaining pure compounds from automated parallel synthesis strategies. One is to optimize the chemistry and synthesis work-up procedures to minimize side products and excess reactants so that the products can meet purity criteria within the reaction well. The second approach is to develop broad-based high throughput purification techniques to separate the product from the by-products and excess reagents present in the reaction mixture. Reverse phase HPLC, which may be viewed as the most sophisticated workup process, is amenable to a large selection of pharmaceutically relevant chemistry space and has become the dominant high throughput purification technique for early drug discovery. Chemistry optimization can be more time consuming and therefore ultimately more costly than purification and has limitations to the type of chemical processes that can be optimized effectively, therefore, this approach is less desirable. In the broad-based HPLC purification approach, products and by-products are separated chromatographically and collected using automatic fraction collection, which is triggered when positive compound identification by mass spectral or UV detection above a minimum threshold has occurred. Moreover, since every analytical characterization method has certain inherent limitations, purity assessment of an optimized crude mixture may only identify those components that are amenable to the analysis technique and thus the true purity would not be determined. By selectively isolating a small fraction of the total chromatographic separation, preparatory HPLC significantly reduces the possibility of having undetectable components in the final sample aliquot. Although high throughput HPLC purification is an important, interesting and a rapidly emerging technique, a more detailed discussion of methods and processes is outside the scope of this discussion.

Assessing the purity of small molecules in a parallel, high throughput environment is not a facile task. Medicinal chemists traditionally utilize NMR and IR to monitor reaction products and yields, but these approaches are time consuming and therefore costly. HPLC has rapidly gained popularity as the principle analytical method for combinatorial or high throughput parallel analysis. However, traditional chromatographic methods are too time consuming to meet the demanding task of analyzing molecule libraries containing hundreds or even thousands of compounds. Moreover, the choice of an appropriate chromatographic detector has become a topic of significant debate. While other detection methods such as refractive index have been used, these techniques do not demonstrate the sensitivity required to detect nanogram-level amounts of impurities.[5] Mass spectrometry has emerged as an important detector in combinatorial library analysis, however, it cannot provide an adequate purity evaluation, since the diverse compounds created within a library may have significant differences in ionization efficiency.[5,6]

Purity evaluation by UV detection is considerably more suitable than the techniques mentioned previously. A drawback of UV detection, however, is that response is dependent upon the presence of chromophoric groups on the compounds to be analyzed. Although shorter wavelengths make the analysis more generic, the response is still dependent on molecular structure and therefore may not be equivalent between starting material, products and by-products. Studies to measure the UV responses of drug-like compounds ranging in molecular weight (MW) from 200 to 700 Da were done by Liling Fang and coworkers.[7] Commercial compounds were measured at a concentration of 100 μg/ml using UV_{214}. The results indicated that higher MW compounds (600–700) exhibited estimates based on the UV absorption much higher in concentrations than the expected 100 μg/ml, while the lower weight (200–300) compounds were estimated at slightly lower concentrations than the actual values.[7] Fang and coworkers concluded that the low-MW compounds have less conjugation resulting in a smaller UV extinction coefficient, while the high-MW compounds exhibit larger extinction coefficients as a result of extended conjugation.

Moreover, in addition to extinction coefficient variations between the components of the synthesis mixture, the absorbance of mobile phase solvent and early elution of sample solvent may interfere with product detection at 214 nm. However, in general this method is applicable to detecting intended products as well as unwanted side-products and un-reacted starting material.

ELSD has been employed for only a few years in support of high throughput synthesis, but has become very popular due to the non-specific nature of the technique. Similarly to the problems discussed above when UV detection was used alone, the purity of small molecule libraries may be overestimated by ELSD due to a reduced response from smaller MW impurities. This phenomenon can be attributed to the fact that ELSD is a mass-sensitive technique that is dependent on the size of a molecule and its vapor pressure.[8] Studies have indicated that low-MW (200–300) compounds were found to have either weak or no ELSD signal, primarily due to evaporation of the compounds with the solvent. This phenomenon has also been observed with final products of low-MW, but more often associated with starting materials identified in the reaction mixture. However, response may be improved with the use of acid modifiers like trifluoroacetic acid (TFA) in the mobile phase, which tend to ion-pair with primary and secondary amines in the reaction mixture forming stable salts that are detected more readily than the MW of these compounds would otherwise suggest. Another potential source of error for ELSD analysis is the inherent non-linearity of the technique, which may overestimate major components at the expense of the minor ones. Although ELSD may at times underestimate the amount of low-MW or concentration of impurities/byproducts, the overall relative variations observed among the large number of different chemistries is less than that observed with UV. In general, detection of starting materials and by-products in combinatorial library analysis is more challenging with the sole use of ELSD.[7] We have found that using both ELSD and UV detection in conjunction with mass spectrometry provides the best means to evaluate library purity. Generally, a purity value of 70–90% as determined by UV or ELSD analyses is acceptable for compounds that will be used for early drug discovery.[5,8]

5   QUANTITY DETERMINATION

The primary compound quantity determination is based on gravimetric analysis of purified compounds because the major issue one faces in quantifying parallel synthesis libraries using an analytical approach is the chemical diversity of the molecules within them and the fact that these libraries contain novel compounds, which are being synthesized for the very first time. For these reasons, it is not possible to quantify these molecules using standard calibration curves, where standards of the product molecule have been synthesized and well characterized. Because of the mass-sensitive nature of ELSD, HPLC/ELSD has been used for the quantification of libraries produced by parallel synthesis. By approximating the response of the detector to be similar for all compounds present in a synthesis mixture, this technique[7–9] can be used as a universal detection technique to analyze diverse library compounds with reasonable accuracy and without using well-characterized individual reference standards. However, studies have shown that optimal quantification accuracy is only achieved with structurally related standards.[8] Selecting standard compounds during library development, for example at the test plate stage, ensures that they contain representative structural functionalities that would be indicative of the entire library. When choosing standards for a particular chemistry, careful consideration should be made with regard to the median MW range, and detector response, as well as the ability to synthesize and purify the compounds.

The ELSD response with respect to concentration is a quadratic function over the entire dynamic range and therefore the expected range of product concentrations must be well represented in the generation of calibration curves.[8,10] A calibration curve over a limited dynamic range using seven calibration standards obtained by serial dilutions is shown in Fig. 5.2 and is indicative of what is used in our process. The expected product concentration is represented as the median value, as well as possible extremes within about one order of magnitude in order to encompass potential concentration variability. Triplicate injections of each dilution are performed to determine statistical variation and evaluate response. The coefficients from a calibration curve may then be used to estimate compound quantity based on area response and therefore accurate peak integration is of paramount importance for quantification. It is recognized that individual compounds may be over or under estimated depending on the similarity of their response compared to the chosen standards, therefore standards should be chosen for their median response to be a reasonable representative for the variety of molecules present in the mixture.

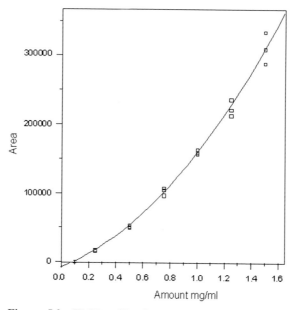

Figure 5.2   ELSD calibration curve generated from a representative library compound; 0.1 – 1.6 mg/ml covers the range of sample concentrations that are expected after synthesis.

Pre-purification quantification may be evaluated using ELSD as part of the initial evaluation of production success in order to determine if purification should be conducted. If the data suggest that the synthesis was successful and enough quantity is available to meet final criteria, the samples are typically diluted with DMSO in preparation for purification. Gravimetric analysis at this point is neither practical nor required. After purification, evaluating quantity by ELSD provides a comparative methodology to gravimetric analysis.

6 FAST CHROMATOGRAPHY

The number and diversity of compounds produced by parallel synthesis approaches makes the cost and time of traditional method development prohibitive. Due to the emphasis on creating a single compound per well in a spatially addressable format, parallel synthesis samples contain a relatively simple mixture when compared to samples from a biological matrix. The parallel synthesis samples typically contain two to five starting materials, which usually have very different characteristics from the corresponding products. Significant resolution is not required, since the purpose of the analysis is to determine if the intended reaction has occurred and how much product has been formed. Fast chromatography, a relatively new approach that focuses on the development of generic methods to analyze many product compounds with "adequate" resolution, is quite amenable to the high throughput separation requirements of parallel synthesis libraries.[4,11-14] Impurities such as reaction by-products and excess reagents are resolved from the products, but their characterization is not emphasized. Additionally, the fewer the number of components in a sample, the lower the number of theoretical peak capacity required for separation and the faster the run time that can be achieved.[4] This is in contrast to more traditional method development where there is usually one target compound and chromatography methods are developed to examine impurities or to quantify the target compound in the presence of complex biological or environmental matrices.

Analytical techniques that can be used to address the questions of purity, quantity and identity in a high throughput approach and that are amenable to automatic and unattended operations are HPLC and MS.[3] In fact, the inherent sensitivity and automation capabilities of these techniques have been the impetus for the advances that allow for the annual characterization of tens of thousands of compounds per system. In addition, scaled-up versions of these techniques have been combined with innovations in semi-preparative chromatography and fraction collection for product purification.[2,15-17] Compound purity, quantity and identity can routinely be determined in about a 2.5-min analysis and similarly a 5-min purification run.

A measure of a column's ability to separate a compound mixture in a gradient approach is peak capacity, which may be defined as the theoretical number of peaks that can be baseline resolved.[4,18] Peak capacity is directly related to the flow rate and column parameters. Short, relatively wide columns, 4.6 mm × 30 mm (3.5 μm particle size), are typically employed for the development of fast chromatographic methods. These dimensions allow for the use of high flow rates, approximately 3–5 ml/min. Conventional column lengths limit the flow rates that may be used due to the excessively high back pressures generated during gradient operation.[2] Previously, Goetzinger et al.,[4] demonstrated that using short columns at high flow rates resulted in high peak capacities and a 5-min method could achieve sufficient resolution for this type of analysis and a 20-min method employing longer columns and more traditional conditions was unnecessary.

Typical reverse phase separations for the elution of compounds with varying properties require gradient elution with gradients implemented over at least 10 column volumes when the retention characteristics of the samples are unknown. The generic methods established for the analysis of parallel synthesis libraries typically employ universal gradients of 10–90% organic solvent in less than 5 min and have been found to provide adequate resolution for many small molecule libraries. At ArQule the generic HPLC methods developed typically employ a gradient of 15–95% organic solvent in less than 2 min. A typical UV and ELSD chromatograms of a three-component HPLC standard obtained with a rapid 2.5-min analysis is shown in Fig. 5.3. These methods provide adequate resolution in a 2.5-min method, which translates to approximately 500 samples that can be analyzed in a 24-h period and the available positions on the injector can hold over 1000 samples, which can easily accommodate an unattended weekend analysis.

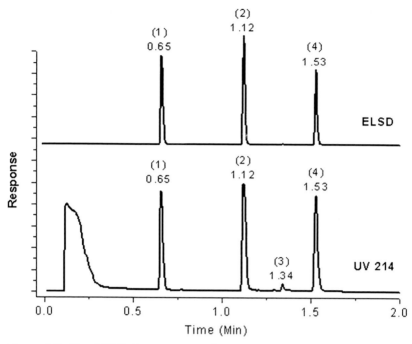

Figure 5.3   Fast HPLC chromatograms of a three component HPLC standard mix (2-hydroxydibenzofuran, 2-acetamidophenol, 3-(4-*tert*-butylphenoxy)benzaldehyde). Flow: 3 ml/min, inj. vol: 3 μl, 0.1% formic acid modifier, Zorbax C8 4.6 mm × 30 mm² column. (a) ELSD trace at 38°C and (b) UV 214 nm trace.

These methods are amenable for use with mass spectrometry as well as UV and ELSD detection and have proven to be reliable for over four years of extensive use.

7   HIGH THROUGHPUT HPLC/MASS SPECTROMETRY

To confirm product identity fast chromatography may be combined with mass spectrometry. MS detection is specific for the determination of MW and the ability to confirm combinatorial synthesis products by logical fragmentation.[19,20]

Reversed phase chromatography is currently the separation mode of choice due to the relative ease of interfacing the mobile phase compositions and sample matrices with mass spectrometry. However, other modes of separation, such as affinity capillary electrophoresis[21] normal phase, ion exchange,[22] and super critical fluid chromatography,[6,23] continue to be developed as potential orthogonal techniques complementing reverse-phase HPLC.

Bench top, single-stage quadrupole mass spectrometers are the detectors of choice for high throughput HPLC analysis because of their relatively low cost, ease of operation, reliability and general ruggedness. These instruments employ electrospray ionization (ESI) and atmospheric pressure chemical ionization (APCI)

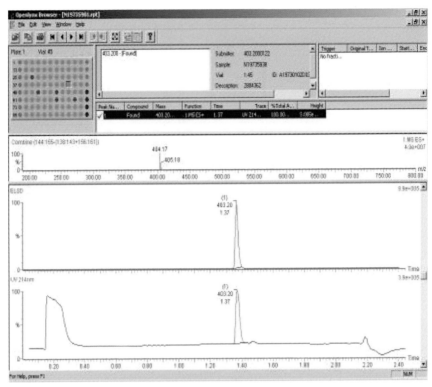

Figure 5.4 Analytical data for typical high throughput analysis of synthesis samples. The window in the left-hand corner shows the pass/fail results of the 96-well plate. The mass spectrum corresponding to the main peak in the ELSD and UV traces are reported in the pane above the analog traces.

probes positioned orthogonally to the sample cone for ease of cleaning and continuously analyzing "dirty" biological or environmental samples at relatively high flow rates.

Mass spectrometry is a gas phase ion-selection technique and therefore, signal depends on the structure of a molecule and its ability to stabilize a charge. Mobile phase modifiers such as formic acid, TFA and ammonium hydroxide may be employed to improve product ionization under ESI or APCI conditions. Both ESI and APcI are amenable to continuous flow LC conditions and are popular for library analyses. Most of the molecules produced in parallel syntheses approaches contain nitrogen, as they are generally modeled after known drugs and natural products and these compounds are amenable to ESI + analyses. Similarly, carboxylic acid containing molecules tend to stabilize negative charges and may be analyzed using ESI- or APcI-. A more detailed discussion of the application of mass spectrometry to combinatorial chemistry is outside the scope of this chapter, however, several excellent reviews have been written on the subject.[6,24–29]

The development of fast LC/UV/ELSD/MS methods has made possible the determination of identity, purity and quantity of parallel synthesis compounds in a high throughput mode compatible with the production of commercial parallel synthesis libraries.[3,10,13,23,30] An example of this type of analysis is displayed in Fig. 5.4, showing a sample page from the MassLynx Browser application. The data were collected in a high throughput manner with method run times of 2.5-min. The top trace displays the mass spectrum of the peak detected by ELSD and UV as seen by the chromatograms beneath it.

8 DATA MANAGEMENT

The advances in chromatography, mass spectrometry and automation have allowed us to provide HPLC/MS analysis for every sample generated by the high throughput parallel synthesis system, as well as virtually every sample generated by development chemistry. Presently, our laboratory analyzes and processes close to a million samples per year using 12 HPLC/MS systems with the expectations to increase to 1.25 million individual samples by the end of 2004. These samples represent a broad spectrum of activities encompassing library protocol development as well as high throughput parallel synthesis, and include early synthesis route scouting, reagent qualification, analytical method development, test plate synthesis, pilot synthesis and production.

The automated high throughput environment of parallel synthesis library analysis creates a huge amount of data. This, in turn, generates an enormous need for automated data management tools.[3] In reality, in order to generate and process this amount of data, the workflow systems within the analytical department must be highly automated and optimized to allow the instruments to run close to the theoretical capacity defined by the sample-to-sample injection time. Moreover, the more automated the transfer and archiving of data can be the less chance for human transcription errors, which compromises data integrity. In addition, by registering compounds in a database such as ISIS and linking the structures and masses of the compounds with the analytical data throughout the synthetic and analytical process, helps identify trends, solve unexpected problems more quickly and increases the overall efficiency of the development and production process.

The most powerful data management activities arise from interfacing and integrating analytical instrumentation with laboratory information management systems (LIMS) databases, as well as other corporate or synthesis databases using web-based tools and custom software applications. For example, if one LIMS is used for storage of analytical results and includes information such as sample ID, peak number, purity, quantity, and detector type and another database is used to store project chemistry information such as monomer lists, templates, and production layouts linked to ISIS, which presents structural information then a software application that allows the viewing of peak purity by production layout provides a powerful way to evaluate reagent success based on chemical structure. Compounds that pass purity criteria may be one color and failed samples another

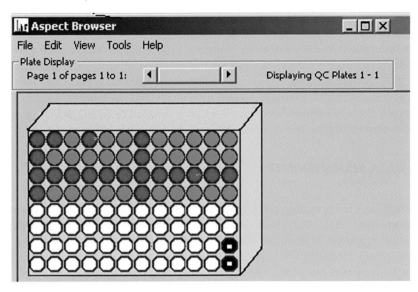

Figure 5.5   Proprietary information management system view indicating which samples meet acceptance criteria. The pattern of dark wells identifies reagents that do not meet acceptance criteria and should be removed prior to library production. The black and white wells represent analytical standards used to evaluate the system's performance. White wells do not contain samples.

creating a pattern for the viewer that summarizes the data. This approach applied at the reagent qualification stage of development can identify a monomer that is known to produce side products based on its "pattern" of failure and subsequently it may be removed from the production plan. An example of such a custom application, which integrates analytical information with production layout, is shown in Fig. 5.5. For this example, two reaction blocks, 4 × 6 wells each, have had samples transferred to a 96-well plate in a side-by-side manner. The reagent in column 1 for both blocks (cols 1 and 7 in diagram) did not produce the desired compound and will be removed from further synthesis. Additionally, the reagent used in row C will be eliminated, as well.

Another application of integrated database tools is analytical data quality control. An analyst can view the data stored in LIMS to ensure its quality, and a summary table forwarded to a different database that can be accessed by other groups. In effect the analyst approves the data and releases it for viewing by non-analytical personnel that may then use the results to make decisions about the process.

The ability to generate large amounts of analytical and synthesis data coupled with the ability to quickly store the pertinent information in the databases provides the means to generate huge databases very quickly. Although these databases contain critical information about the products and processes, every group within the company has specific needs for information depending on their specific function. Therefore, effective software tools that help summarize and manage the flow of information between and among work groups are just as important as the databases themselves. Custom reports that allow a user to track and evaluate the progress of an array through the process are very helpful. For example, the number of development plates, reagent qualifications and production runs that were executed and the corresponding analyses and approval dates may be summarized and sent to the person responsible for a particular chemistry. Other software tools that allow the user to examine compound structures within the context of a plate submitted for characterization allowing chromatographic results to be correlated to specific reagents are also available and quiet useful. Interactive software that can search across various databases and rapidly present information necessary to make decisions is critical to the company's success.

9   EXPERIMENTAL METHODS

Autosamplers capable of storing up to twenty-four 96-well plates, fast chromatography, rapid MS techniques with multi-user interfaces, and improved data analysis/storage systems make up the increasingly powerful arsenal of analytical tools. The key to any successful analytical process is integration of the individual components. Ideally, only one software package would be needed to control the autosampler, HPLC and MS instrument parameters. In reality, there are numerous solutions to this problem ranging from contact closure at the most basic level to highly integrated systems. The following descriptions are of the components of the system currently in use at ArQule.

9.1   Sample Introduction

Samples are injected using LEAP autosamplers to maximize throughput. The cycle time of the injector is short enough that the injector can be used to run a 1.0-min method. Plate storage consists of two stacks of drawers. One stack contains three drawers that can accommodate deep well plates and the other stack contains six drawers that can hold shallow well plates. Each drawer is deep enough to hold two plates and a side rack capable of holding two additional plates or vial racks allows for a total of 20 plates or 18 plates and two vial racks. Accurate, reproducible, low volume sampling can be achieved by using a 10 μl syringe.

The Waters MassLynx software allows for control of the LEAP through the Inlet Editor and allows for the creation of customized sampling cycles. Carryover was eliminated by using the available programming of the LEAP injector and programming the syringe to rinse within the wash port and rinse the injection port.

9.2   HPLC Instrumentation

Characterization analyses were performed using a Shimadzu LC-10AD VP binary pump system with an SCL-10A VP system controller. Method conditions were established on the system controller and start run was initiated by contact closure through the MassLynx software of the Waters spectrometer. A Shimadzu SPD-10A VP variable wavelength UV detector was used to monitor 214 and 254 nm to assess purity. The mobile phase was split from the chromatographic column post-UV so that approximately 100–200 ul continues on to the mass spectrometer. The remaining flow is directed to an ELSD detector. The columns used are Zorbax SB C8 or XTerra MS C8, 4.6 mm × 30 mm, 3.5 μm particle size, depending on the mobile phase modifier and pH.

9.3   MS Instrumentation

Waters ZQ and ZMD single quadrupole benchtop instruments are used in our laboratory. The methods of choice are ESI and APcI. Under ESI conditions the cone voltage is set to 5–20 kV, and the source temperature from 125 to 150°C, with desolvation temperatures around 250°C. Typical APcI conditions employ a corona voltage of 2–4 μA, cone voltage of 45 V and probe temperature of 500°C. MassLynx 3.5 software is used to acquire and process sample data.

9.4   Analytical Standards

DMSO blanks, HPLC and MS analytical standards are placed into each 96-well plate to monitor the performance of the instrumentation during a run. For an acid analysis, the HPLC standard is a mixture of 2-acetamidophenol, 3-(4-*tert*-butyl-phenoxy)benzaldehyde, and 2-hydroxydibenzofuran at concentrations of 1 mg/ml in DMSO. These compounds respond well by UV and ELSD and cover the elution range of the gradient under the conditions of the analytical method being applied. The performance of the HPLC instrumentation and columns are evaluated using this standard. The MS standard is a mixture of chlortetracycline hydrochloride, quercetin dehydrate, rhodamine B and flavone in DMSO. The MS standard is used to monitor mass accuracy and ionization performance. In typical analytical runs, analytical standards bracket the sample plates and the results obtained prior to and after each plate are compared in order to verify that all instrumentation is operating according to specifications and the data generated is valid and consistent throughout the run.

9.5   ELSD Quantity Determination

The quantities of individual compounds within a library are determined using an ELSD detector to generate a response *vs* concentration calibration curve. Compounds that have been identified as characteristic of the expected library products are used as analytical standards for the rest of the library members. Calibration curves are generated using seven concentrations (0.1, 0.25, 0.5, 0.75, 1.0, 1.25, and 1.5 mg/ml) bracketing the expected theoretical concentration of the sample obtained by serial dilutions of a 3 mg/ml stock solution. Each standard is analyzed in triplicate and the best quadratic function is fit to the response *vs* concentration data. Library sample concentrations are then calculated using the identified peak area and the equation for the calibration curve. The ELSD quantitative method may be used to validate the final product quantity, which is determined gravimetrically and is adjusted based on ELSD and/or UV purity.

9.6 Software

HPLC/MS data capture and analysis was performed using the Waters/MassLynx (versions 3.4 and 3.5) software and the Justice Laboratory Software ChromPerfect Spirit. Although the MassLynx software acquires and presents the mass spectrometric as well as the UV and ELSD data in an integrated format, the software is primarily geared for mass spectrometric data. Thus, there is minimum digitization of the analog traces. For most conventional separations where UV/ELSD peaks are 30–40 s wide sufficient points are acquired across the peak to adequately define its shape and size. However, with our fast chromatography application, which generates analog peaks that can be less than 2 s wide, the MassLynx (v3.4/3.5) digitization rate was not enough to adequately define the size and shape of these peaks. In order to more accurately represent the inherent resolution of the separation method by the analog traces, a faster chromatographic data acquisition system was necessary. The ChromPerfect Spirit system was used in parallel with the MassLynx system. The MassLynx MS data was used for identifying and confirming the peak associated with the compound of interest, whereas the purity information for each sample was determined from the corresponding peak obtained from the ChromPerfect system. The correlation between data systems was performed by custom software applications and the LIMS database.

10 CONCLUSION

The need for more efficient and effective compound synthesis to support drug discovery has fueled the development of high throughput parallel synthesis as an opportunity to expand the efficiency of traditional medicinal chemistry. However, the success of the automated synthesis platform is not only dependent on the efficiency of the synthesis process and the broad chemistry that can be performed but also on the proper characterization of the produced compounds. As a result, a high throughput rapid HPLC/MS analysis has been developed that has required advances in chromatography, data acquisition and management, as well as significant process and workflow efficiencies.

REFERENCES

1. F. Balkenhohl, *Angew. Chem. Int. Ed. Engl.*, **35**, 2289 (1996).
2. H.N. Weller, *Mol. Divers.*, **3**(1), 61 (1997).
3. J.N. Kyranos and J.J. Hogan, *Anal. Chem.*, **70**(11), 389 (1998).

4. W.K. Goetzinger and J.N. Kyranos, *Am. Lab.*, **30**(8), 27 (1998).
5. B.H. Hsu, *J. Chromatogr. B, Biomed. Sci. Appl.*, **725**(1), 103 (1999).
6. X. Cheng and J. Hochlowski, *Anal. Chem.*, **74**(12), 2679 (2002).
7. L. Fang, *J. Comb. Chem.*, **2**(3), 254 (2000).
8. C.E. Kibbey, *Mol. Divers.*, **1**(4), 247 (1996).
9. N. Shah, *J. Comb. Chem.*, **2**(5), 453 (2000).
10. L. Fang, J. Pan, and B. Yan, *Biotechnol. Bioeng.*, **71**(2), 162 (2000).
11. B.D. Dulery, J. Verne-Mismer, E. Wolf, C. Kujel, and L. Van Hijfte, *J. Chromatogr. B, Biomed. Sci. Appl.*, **725**(1), 39 (1999).
12. I. Hughes and D. Hunter, *Curr. Opin. Chem. Biol.*, **5**(3), 243 (2001).
13. J.N. Kyranos, *Curr. Opin. Biotechnol.*, **12**(1), 105 (2001).
14. F. Leroy, *J. Chromatogr. Sci.*, **39**(11), 487 (2001).
15. L. Zeng, *Comb. Chem. High Throughput Screen.*, **1**(2), 101 (1998).
16. H. Cai, *Rapid Commun. Mass Spectrom.*, **16**(6), 544 (2002).
17. H.N. Weller, *Mol. Divers.*, **4**(1), 47 (1998).
18. L.R. Snyder, J.J. Kirkland, and J.L. Glajch, *Practical HPLC Method Development*, 2nd ed., Wiley, New York (1997).
19. A.S. Fang, *Comb. Chem. High Throughput Screen.*, **1**(1), 23 (1998).
20. Y. Dunayevskiy, *Anal. Chem.*, **67**(17), 2906 (1995).
21. Y.-H. Chu, *J. Am. Chem. Soc.*, **118**, 7827 (1996).
22. M.A. Strege, *Anal. Chem.*, **72**(19), 4629 (2000).
23. J.N. Kyranos, *Curr. Opin. Drug Discov. Dev.*, **4**(6), 719 (2001).
24. R.D. Sussmuth and G. Jung, *J. Chromatogr. B, Biomed. Sci. Appl.*, **725**(1), 49 (1999).
25. C. Enjalbal, *Mass Spectrom.*, **19**(3), 139 (2000).
26. G. Siuzdak and J.K. Lewis, *Biotechnol. Bioeng.*, **61**(2), 127 (1998).
27. C. Enjalbal, *Comb. Chem. High Throughput Screen.*, **4**(4), 363 (2001).
28. A. Triolo, *J. Mass Spectrom.*, **36**(12), 1249 (2001).
29. D.B. Kassel, *Chem. Rev.*, **101**(2), 255 (2001).
30. L. Fang, *Rapid Commun. Mass Spectrom.*, **16**(15), 1440 (2002).

6

High Throughput Parallel LC/MS/ELSD of Combinatorial Libraries Using the Eight-Channel LCT System with MUX Technology

Peter W. Davis

Nereus Pharmaceuticals Inc., 10480 Wateridge Circle, San Diego, CA 92121, USA

Michael C. Griffith

University of California, San Diego UCSD Extension Bioscience, 9500 Gilman Drive, MC 0172S, La Jolla, CA 92093-0172S USA

CONTENTS

1 INTRODUCTION

The objective of our company, Trega Biosciences, was to create novel druglike compounds for "hit" discovery. Solid-phase organic synthesis methods were used to make large arrays of single compounds. A typical synthesis set yielded about 8000 compounds on a scale of ~10 mg of each. Intermediate compounds were analyzed for identity and purity prior to final reactions, cleavage, and separation

High Throughput Analysis for Early Drug Discovery
Edited by James N. Kyranos

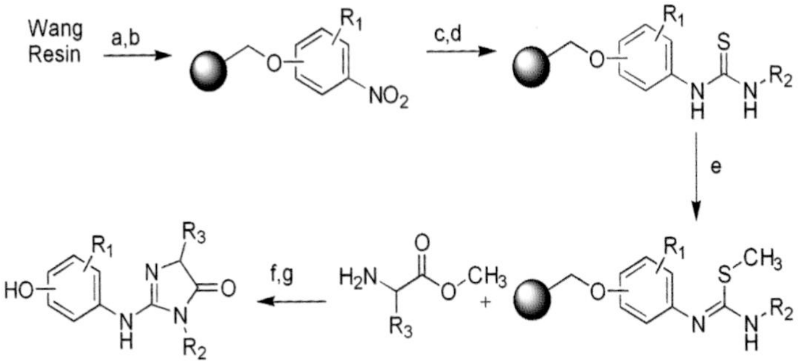

Figure 6.1 Synthesis scheme for a typical set of library compounds.

from the resin. Only intermediates of approximately 75% purity were used to make the final array of compounds. However, this was not a guarantee of successful synthesis of the final products, so the final compounds were analyzed by high performance liquid chromatography/mass spectrometry/evaporative light scattering detection (HPLC/MS/ELSD) for unambiguous determination of identity and purity. Figure 6.1 shows the synthesis scheme for a typical set of library compounds.

Our chemistry laboratories produced approximately 200 000 compounds per year, requiring analysis of an average of about 4000 compounds per week. In addition, we were interested in routine re-analysis of libraries to determine the stability and quality of compounds over time. The large numbers of required analyses necessitated either several systems, or a single system capable of the throughput of several conventional systems. This chapter describes the use of the commercially available MUX-LCT (MicroMass) system to analyze eight HPLC separations simultaneously. The system allowed us to meet our goal of analyzing approximately 300 000 compounds per year.

Since many laboratories require high quality data on a limited budget, it is important that not only the equipment be affordable, but that the downstream data handling does not require a large investment of time or money. We demonstrate that fairly simple routines can be used to process, present, and store results of analyses. The results were also summarized in a format useful for input into the widely used ISIS database (MDL).

2 EXPERIMENTAL

Compounds were synthesized in 96-well plate format on solid phase. For screening purposes the first and last columns were left empty, resulting in a total of 80 samples per plate. The compounds were cleaved from the resin using trifluoroacetic acid and the solution was pipetted directly into fresh plates after settling of the resin to the bottom of the well. The solvent was removed under reduced pressure in a vacuum centrifuge (Genevac), and the products were resuspended in acetic acid.

Typically 5 μl of this acetic acid solution was diluted into dimethyl sulfoxide (DMSO) in an analysis plate to a final volume of 150 μl for an approximate concentration of 1 μg/μl. A 35 μl injection through a 5 μl loop resulted in an analysis of approximately 5 μg of crude sample.

The LC/MS system consisted of an LC pump (Waters) with the flow split into eight lines feeding into an eight-channel liquid handler (Gilson) and then to eight HPLC columns. After separation by HPLC, the flow was further split to one of eight SEDEX ELS detectors (Sedere) and one of eight inlet ports of a MUX multiplexed ESI source interfaced to the LCT time-of-flight (TOF) mass spectrometer. A Plate Crane (Hudson Control) was used to load plates onto the Gilson bed. The HPLC method was typically a 6-min linear gradient of 0.05% TFA/acetonitrile in 0.05% TFA/H$_2$O (10% aqueous to 100% organic) through Phenomenex Synergy 2 mm × 50 mm 3-μm C18 columns fitted with 0.5-μM inline filters (MetaChem). The flow across each column was nominally 1.0 ml/min; the split results in about 75 μl/min to the MS and the balance to the ELSD. Figure 6.2 shows a schematic of the eight-channel LC TOF MUX/ELSD system.

Total Control software provided with the Plate Crane controlled the MassLynx software *via* the AutoLynx feature. This automation feature also allowed concurrent processing of the data and "printing" to a text file for further processing in Excel. Data were processed offline on a second PC workstation. Each library produced approximately 2–3 GB of data, so a 5.2 GB DVD-R drive was used to back up the data as the PC hard drive filled.

System suitability was routinely monitored using a solution of 0.5 mg/ml glipizide in DMSO. Glipizide is similar to the types of compounds contained in our libraries. It is stable for extended periods in DMSO solution at room temperature. Calibration of the ELSDs was also performed with a serial dilution of glipizide in DMSO. This calibration tended to hold well for several days, unless there was a

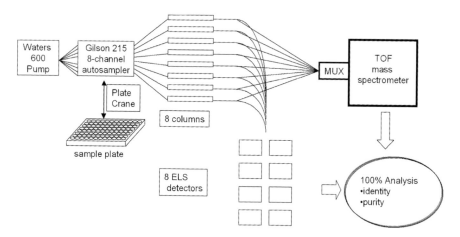

Figure 6.2   Schematic of the eight-channel LC/LCT-MUX/ELSD system.

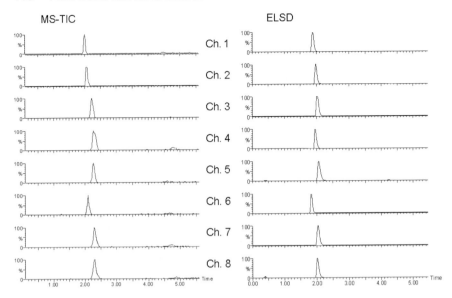

Figure 6.3   Typical system suitability analysis with glipizide.

drastic change in the system such as a new ELSD lamp or a change in the mobile phase. Figure 6.3 shows a typical system suitability analysis with glipizide.

3   RESULTS AND DISCUSSION

Initial testing for a new compound library was performed on four plates (totaling 320 compounds) to adjust gradients and MS parameters. Typical compound sets were easily analyzed using general HPLC and MS methods. In the case of particularly hydrophobic or hydrophilic compounds, the gradient was adjusted to ensure elution of product well within the observable time range. Adjustment of the gradient was straightforward and usually completed with the test plates. In some cases, significant fragmentation of the products or very poor ionization during MS analysis was revealed during the initial plate analysis. Optimization of the MS method required careful investigation of parameters, and was best performed, if possible, with a sample of exemplary product procured from the chemist, so as not to waste any of the actual library compounds. Some examples of difficult compound sets include phenolic and benzylic ethers, which tended to fragment very readily, and tertiary sulfonamides, which tended to ionize very poorly. Threshold levels for the acceptable percentage of the base peak showing the expected mass were set quite low, typically 5%. The background on the LCT is very low and false positives were rarely observed.

   For purity analysis we chose ELS detection, perhaps the only detector other than UV that is universally accepted. While both have their positives and negatives, we believe that ELSD is more accurate for our application. Unlike UV detection, the signal is not sensitive to the chemical nature of the chromophores in the product and impurities. For compounds within a general class and molecular weight, the ELSD better represents the relative mass of materials present. Solvents such as DMSO give a large peak in the low-wavelength UV, but are essentially invisible to the ELSD. One limitation of the ELS detectors is their limited dynamic range. Typical UV detectors have a range of two orders of magnitude or more. The ELSD, however, has a useful range of just over one order of magnitude. Thus, it can be difficult to determine the purity of collections of samples with widely different concentrations, as some peaks will be off-scale while others will be too small to measure. In addition, small levels of impurities tend to not be observable in the chromatogram.

The analysis software, OpenLynx, successfully correlates analog peaks from ELS or UV data with mass spectra; however, further processing is often required depending on the user's needs. Complete sample tracking and data processing has been described by researchers at Amgen.[1] Although the number of samples we were processing was high, the downstream data handling need not be sophisticated, and so the expense of a high-level data handling solution was not warranted. We found that we were able to process the "results summary" text file generated by the OpenLynx using Microsoft Excel macros written internally.

Sets of data each representing one plate of compounds were imported into Excel, reduced to a single line per compound where appropriate, and the results were summarized in various degrees of detail, as shown in Fig. 6.4. In this example, OpenLynx identified the ELSD peaks as product (Fig. 6.4a), and further processing determined the ELSD-based purity of the product peaks and whether or not the well passed chosen purity criteria. Figure 6.4b shows the format used to enter the processed data into our ISIS database.

It is worth noting that there are several peaks identified as product in the example shown in Fig. 6.4a. In this case, we determined that for this compound set there should be only a single product, so only the largest peak was used to determine the purity. We have analyzed libraries containing diastereomers that were resolved using the HPLC method. Another set of macros was used to allow multiple peaks to be combined to generate the overall purity. The form of the output allows very flexible processing of the results, and other users may find that only modest experience writing Excel macros will be sufficient to generate similar automatic processing of results.

Figure 6.5 represents further processing of the plates into plate summary data (Fig. 6.5a) and a visual representation of the analog results (Fig. 6.5b). These summary representations are useful for discerning patterns in the analytical results, as illustrated in Fig. 6.5b. In the final reaction used to synthesize these compounds, each half of the plate was reacted with one of 40 unique reagents, thus wells B3 and B8 in each plate had the same final reaction conditions and reagents. Eight of the failed compounds were in the same position on each of the two plates, leading us to conclude that a systematic failure was occurring due to four of the final reagents used.

File	Date	Time	Target	Found	Trace	Time	Peak	Area %Total	Area Abs
_001-A02	12-Mar-01	14:04:30	496.32	YES	Channel 1	4.09	1	100	183468.1
			496.32	YES	TOF MS ES+ :TIC	4.14	2	100	357.1
			496.32	YES	TOF MS ES+ :497.719	4.14	2	100	286.5
_001-B02	12-Mar-01	14:04:31	498.33	YES	Channel 1	3.75	1	95.34	153969.9
			498.33	YES	TOF MS ES+ :TIC	3.8	2	100	459.6
			498.33	YES	TOF MS ES+ :499.735	3.8	2	100	333.6
			498.33	YES	Channel 1	4.88	3	1.97	3181.4
			498.33	YES	Channel 1	5.05	4	2.69	4347.9
_001-C02	12-Mar-01	14:04:31	482.29	YES	Channel 1	3.69	1	98.33	329215.7
			482.29	YES	TOF MS ES+ :TIC	3.69	2	100	500.6
			482.29	YES	TOF MS ES+ :483.692	3.69	2	100	379.6
			482.29	YES	Channel 1	5.06	4	1.67	5602.2
_001-D02	12-Mar-01	14:04:32	495.29	YES	Channel 1	3.64	1	6.93	13944.2
			495.29	YES	TOF MS ES+ :TIC	3.8	2	100	280.3
			495.29	YES	TOF MS ES+ :496.691	3.8	2	100	224.1
			495.29	YES	Channel 1	3.82	3	72.17	145207.4
_001-E02	12-Mar-01	14:04:32	468.27	NO					

a

Plate Name	Well	ELSD Purity	Pass/ Fail	Plate Purity	91.19
_001	A02	100	1		
_001	B02	95.34	1		
_001	C02	98.33	1		
_001	D02	72.17	1		
_001	E02		0		
_001	F02	89.02	1		
_001	G02	87.67	1		
_001	H02	93.84	1		
_001	A03	97.15	1		
_001	B03		0		
_001	C03	98.89	1		

b

Figure 6.4   Representative data analysis. (a) Results summary text file read into Excel. (b) Results summary reduced to a single line per sample for database import.

Robustness and reliability are critical for a high throughput system, in order to get the most data with the least amount of personnel. While our system was able to analyze nearly what we expected in quality and numbers, there were issues that resulted in excessive "hands on" time. Multiplexed LC/MS is very new technology[2] so we anticipated issues related to the MUX portion of the system. In general, however, we had no real issues with respect to crosstalk between channels or data partitioning. There was also some skepticism about being limited to electrospray ionization, but we found it to be suitable for all of the compound types that we created.

As with typical single-channel LC/MS systems, sample quality affected the reliability of the system. The compounds were crude, unfiltered samples directly from solid-phase synthesis and cleavage reactions. Although the synthesis process results in generally very good yield and purity of compounds, there is the

Plate Number	#of Hit	Pass/ Fail	Plate Purity	Passing Wells	Overall Purity
_001	71	1	91.19	5645	90.91
_002	67	1	89.89		
_003	69	1	92.63		
_004	70	1	88.89		
_005	70	1	91.14		
_006	71	1	91.24		
_007	69	1	85.07		
_008	68	1	85.01		
_009	67	1	89.59		
_010	64	1	83.99		
_011	71	1	90.76		

a

Purity Criteria	50											
Plate _001		2	3	4	5	6	7	8	9	10	11	Failed Wells
	A	100	97	89	100	97	100	95	90	100	97	9
	B	95	0	97	100	0	93	0	99	100	0	
	C	98	99	57	100	100	100	98	53	100	100	
	D	72	45	88	88	97	59	37	87	86	97	
	E	0	100	100	84	56	0	100	100	84	50	
	F	89	99	99	87	93	86	98	97	79	90	
	G	88	100	97	98	100	84	100	96	98	99	
	H	94	88	92	87	84	73	86	79	83	81	
_002		2	3	4	5	6	7	8	9	10	11	Failed Wells
	A	96	91	84	96	96	100	96	86	100	97	13
	B	88	0	97	96	0	96	0	100	100	0	
	C	93	89	28	95	95	100	99	30	100	100	
	D	52	34	87	79	88	63	43	92	89	95	
	E	0	98	99	97	38	0	100	100	90	65	
	F	79	97	92	73	91	91	98	98	88	95	
	G	83	100	98	95	92	86	100	100	99	100	
	H	34	59	50	57	41	81	88	80	88	81	

b

Figure 6.5   Plate data analysis. (a) Summary of results for each plate. (b) Visual representation of results based on purity criteria.

opportunity for minor amounts of insoluble materials to be present in the final solutions. Insoluble matter affected the robustness of the system, and presented challenges in maintenance.

One issue was the robustness of the Gilson liquid handler. These liquid handlers are ubiquitous in robotic analysis systems, due to their reliability and ease of use with a variety of systems. However, they do need maintenance, and this maintenance increases with the decrease in sample homogeneity, due to wearing of seals and clogging of narrow lines. The valve module block of the eight-channel Gilson 215 is fairly crowded (containing eight separate valves) and thus not easy to work on, so if there is a problem it can result in significant downtime. Improving the sample quality either through filtration or SPE treatment could significantly reduce the autosampler maintenance required. Conversely, an alternative autosampler that is easier to work on might be preferred if and when they become available.

Once samples are introduced, there is the possibility of clogging the columns as well. Disposable "javeline type" 0.5-μm filters work very well to trap insolubles and extend column life. A column that exhibits impeded flow nearly always can be fixed with a filter change. We were surprised to find that one column with impeded flow did not significantly affect the performance of the other columns, and would usually yield acceptable data before losing performance altogether. Unfortunately, the parallel nature of the system makes it impractical to re-analyze a single row of data, and it was usually easier to re-run all the rows when the data from one or two was compromised.

In general, many problems were averted by two daily processes. The first was daily running of 100% organic mobile phase through the system for about 30 min. Although all of our gradients included a brief period of flushing with 100% organic phase, this extended treatment reduced regular maintenance of nearly all the components, including the autosampler, columns, ELSDs, and the MS. Another important process was daily analysis of the suitability standard, in our case, glipizide. Since lines seldom went bad quickly, but rather decayed in performance over time, daily tests helped to identify problems and initiate preventative maintenance before data collection was compromised. The most common observations were an increase in retention time due to column or line obstructions, increase in peak width due to column deterioration, and a decrease in MS or ELSD signal due to nozzle clogging. The daily glipizide analysis also allowed a check of the ELSD calibration between complete calibration curves.

Figure 6.3, as noted before, shows typical raw data for a system suitability test during analysis. Minor differences in the flow through each separate column resulted in varying retention times. The variability in peak widths is due to some degradation in performance; the apparent discrepancy between the analog and MS channels suggests the individual detectors (for example, the ELSD nebulizer or MS nozzle), rather than the columns, were responsible. In practice this level of performance is relatively typical, and still acceptable, after running several thousand samples.

When we were using only the Gilson plate bed for sample introduction, we found that large sample queues (for example, 10 plates of 80 samples each) led to system instability. We routinely used a sample queue of 800 samples per run (10 plates of 80 samples per plate) and the system often stalled at random times. When we implemented the Plate Crane, the frequency of this "glitch" dropped dramatically. Our conclusion was that the MassLynx operating software did not deal with large queues effectively. Since the Plate Crane's Total Control software queued only one plate of samples at a time the overall system became much more reliable.

The same features that make ELS detection useful for purity determination over a broad spectrum of compounds also makes it attractive for determining accurate

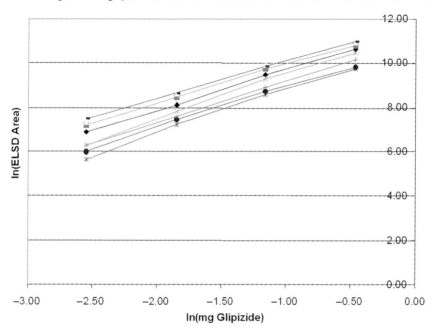

Figure 6.6   Calibration curves for eight channels using glipizide.

amounts of product present.[3] We, therefore decided to evaluate the eight-channel system to see if we could exploit this feature as well. Figure 6.6 shows calibration curves using glipizide at four concentrations spanning the dynamic range of the ELS detectors (see previous comments). We evaluated four different drug compounds (Fig. 6.7) structurally similar to the compounds typically synthesized in our combinatorial libraries, and found that all gave very similar responses, so glipizide was chosen as our standard. Plots of ln (mg) *vs* ln (peak area) were both linear and reproducible. Changes in the calibration were minimal over the course of a few days, and generally only changed significantly due to abrupt changes in lamp performance (including changing a lamp) and nebulizer performance.

Figure 6.8 shows the results of analysis of these four drugs at various concentrations. The error was quite modest, rarely exceeding 20% and typically closer to 5% or less. From this result we concluded that routine calibration of the ELSDs would allow us to estimate the amount of product present in each well. In fact these estimates correlate well with product yield after purification by HPLC (data not shown).

4   FUTURE TRENDS

We believe that the MUX system could take better advantage of the utility of both the TOF-MS and ELSD. For example, the system shows promise for

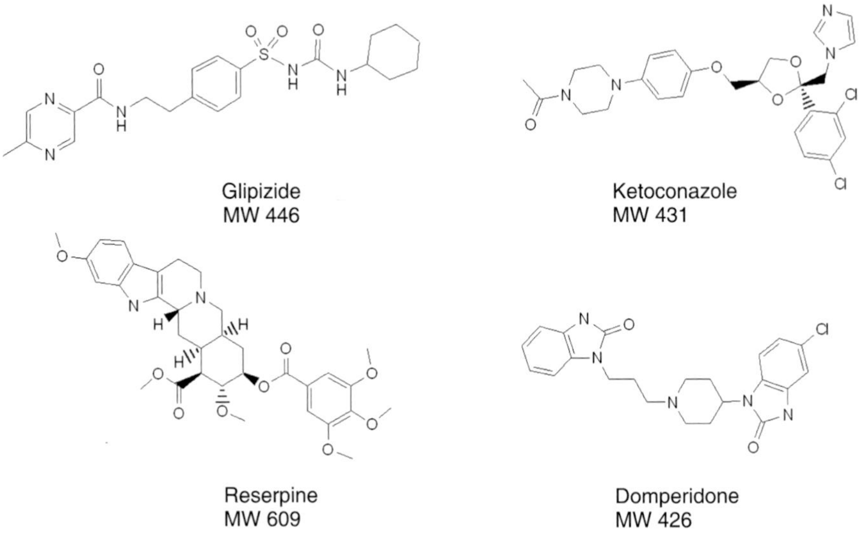

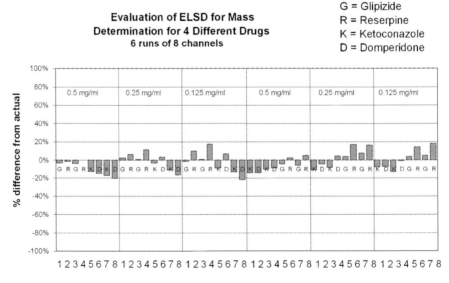

Glipizide
MW 446

Ketoconazole
MW 431

Reserpine
MW 609

Domperidone
MW 426

Figure 6.7   Compounds used for ELSD quantitation testing.

high throughput, quantitative MS analysis of large numbers of samples for pharmacokinetic studies. We also expect that ELSD data will be routinely used as a quantitative tool for estimating the amount of product present in each well. As proteomics grows, the MUX could also become invaluable for keeping up with the growing demand for protein sample analysis.

Figure 6.8   Determination of material present for four drug compounds.

5   CONCLUSION

The MUX-LCT system proved to be fairly robust and allowed us to collect LC/MS/ELSD data on many more compounds than conventional single-channel systems would allow. Under the best of circumstances the system was able to process as many as 7000 compounds per week. The overall throughput was, however, often reduced due to sample characteristics (hydrophobicity, ionizability, fragmentation) and system downtime. The data was easily processed automatically without sophisticated additional software through the use of Excel macros. The system was indispensable for the numbers of compounds we were required to analyze. As this technology matures, better peripherals such as pumps and autosamplers will improve the reliability and the performance of this and other systems, making them true high throughput workhorses.

ACKNOWLEDGEMENTS

The authors thank Dr Pei Chen for developing Excel macros and for helpful suggestions, Chris O'Neill for the initial system setup and operation, and Dr Normand Hebert for providing Fig. 6.1.

REFERENCES

1. S. Chatwin, D.T. Chow, E. Maliski, W. Talen, G. Woo, Y. Zhao, and D. Semin, *Drug Discov. Dev.*, **5**, 50–52 (2001).
2. T. Wang, L. Zeng, J. Cohen, and D.B. Kassel, *Comb. Chem. High Throughput Screen.*, **2**, 327–334 (1999).
3. L. Fang, M. Wan, M. Pennacchio, and J. Pan, *J. Comb. Chem.*, **2**, 254–257 (2000).

7

Purification and Analysis of Parallel Libraries

Cheryl D. Garr

Albany Molecular Research, Inc., 18804 North Creek Parkway, Bothell, WA 98011, USA

Lauri Schultz

CEPTYR, Inc., 3830 Monte Villa Parkway Suite 200, Bothell, WA 98021, USA

Lynn M. Cameron

Applied Biosystems, 850 Lincoln Centre Drive, Foster City, CA 94404, USA

CONTENTS

High Throughput Analysis for Early Drug Discovery
Edited by James N. Kyranos

1 INTRODUCTION

A medicinal chemist, using traditional synthetic organic methods, can prepare approximately two–three purified compounds per week. It often takes 10 years or more to bring a new drug to market, so this approach can be time-consuming and expensive. The advent of combinatorial chemistry has provided a more rapid method to identify and develop drug leads by allowing high throughput syntheses of thousands of reaction products on a weekly basis. High throughput and high-density screening has readily accommodated this large number of compounds. Unfortunately, this creates a new bottleneck where, once hits are identified, the active component of a reaction mixture must be isolated to allow confirmation of its structure and activity.

To obtain maximum usefulness of the wealth of biological information provided by modern screening, it is critical to assay reaction products of known quantity and purity. This can significantly reduce the time and effort invested in following up hits by decreasing the number of false positives and negatives, providing reliable structure activity relationship (SAR) data, and reducing the time required for the isolation and characterization of trace side-products. To this end a process[1] has been developed that includes methods for high throughput preparative scale purification and identification of not only designed products, but also side-products, intermediates and reagents of combinatorial libraries. This method is flexible and can be readily adapted for the isolation and quantification of purified reaction products of both large lead finding and smaller focused combinatorial libraries. This work was originally performed by the Drug Discovery division of MDS Panlabs, currently Albany Molecular Research Bothell Research Center.

2 CHALLENGE

The goal is the large-scale purification of hundreds of samples and analysis with partial characterization of thousands of fractions on a daily basis. The purpose is not the fast turn-around of samples, but rather high throughput; to process as many reaction products as possible, not necessarily as quickly as possible. A modular process was adopted to allow maximum efficiency. If one step in the process becomes inoperable, the remaining steps continue operations thus reducing the impact on production. Further, bottlenecks in the process can be addressed and optimized. For example, if obtaining mass spectral data is a rate-determining step, mass spectrometers can be run for extended hours or additional instruments can be brought on-line. The process described below allows 464 samples to undergo HPLC fractionation on a daily basis. Averaging five isolates per reaction product produces approximately 2320 fractions to undergo downstream processing.

One critical requirement of any multi-step, high numbers process is the tracking of samples. The modular system described must track many manipulations and record the history of isolates as they progress through the purification and identification process (Fig. 7.1). Therefore, in addition to the process itself, a custom compound tracking system (CTS) was developed to meet this need.

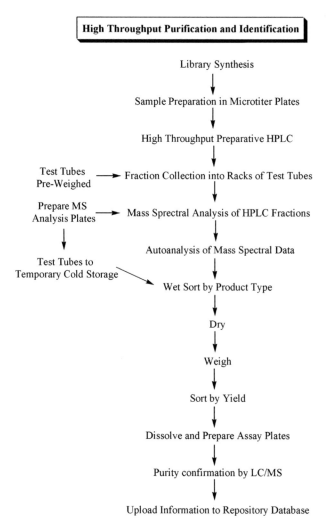

Figure 7.1 Schematic flow chart of high throughput purification process.

Data are stored and tracked on a relational database that is housed on a central server. This allows the status of samples to be continually tracked by the database. Pertinent files are uploaded and downloaded in the content and format required by individual workstations. The data generated as output from one workstation are processed by the CTS to produce input files for the subsequent workstations. Various "if then" scenarios are built into the system, with user-interface options. Reports are generated allowing the progress and end products to be assessed and summarized.

Another key step in the process is the prioritization of samples. The analysis of mass spectral data on thousands of fractions on a daily basis is a daunting task. In keeping with a modular process, mass spectral data is obtained and analyzed in steps separate from peak collection. The data are auto-analyzed in two steps. First, the data are summarized in a spreadsheet that identifies products, intermediates and reagents along with information about their mass spectra. Second, the summary is uploaded into the CTS where compounds are assigned a status: designed product, by-product, recovery item or waste.

The following sections provide a description of each step in this purification and identification process as well as the CTS and prioritization process.

3 DISCUSSION

3.1 Purification

Originally, combinatorial libraries applied to this process were synthesized on a 1 mmol scale using standard solution-phase parallel techniques. As each reaction product is synthesized an aliquot of 0.25-mmol (ca. 100-mg, theoretical) is placed in an individual well of a 48-well microtiter plate for HPLC purification. The remainder of the reaction product is divided equally between three 1-dram vials and placed in $-20°C$ storage. Since original development, this process has been applied to a variety of library types, sizes, and scales. For the purpose of continuity, original protocols will be discussed.

The first step in the purification process is to obtain tare weights on vessels that will be used to collect HPLC fractions. Test tubes are manually labeled and placed into custom racks (Marsh Biomedical Products, Rochester, NY). Each rack has a microtiter plate footprint that allows utilization of existing automation for sample manipulation. Fifteen racks of 24 test tubes can be placed on the deck of a weigh station (Mettler-Toledo Bohdan, Inc., Vernon Hills, IL). The tubes are moved from the racks to a scale and back to the racks using a three-pronged gripper arm. The weigh station automatically tares the balance, scans the barcode, and measures the weight of each test tube. The 24 test tubes in each Marsh rack are further associated with the barcode on that rack.

The second step in the purification process is high throughput preparative HPLC. To achieve the desired throughput and meet the needs of our process, the instrument chosen was the Parallex™ HPLC (Biotage Inc, Charlottesville, VA, Fig. 7.2).

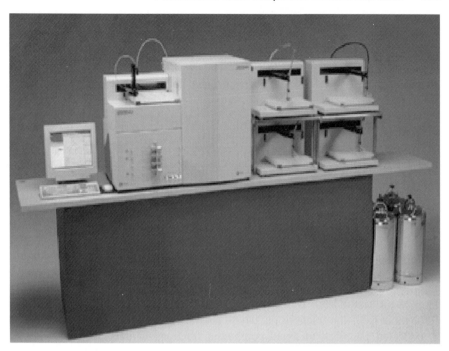

Figure 7.2 Biotage Parallex™ preparative HPLC system.

After initial installation, we became co-developers and a beta-site tester for the system. The instrument fractionates four reaction products in parallel. After all injection loops have been loaded, the contents are injected onto four separate columns simultaneously. The same gradient is applied across all four columns with each column having a dedicated flow path through a dual wavelength UV detector and fraction collector. Injection loops are loaded sequentially while the previous HPLC run is in progress to eliminate down-time between reaction product separations. Using aggressive HPLC conditions, 464 samples, each weighing about 100 mg, can be purified in two 4 h 20 min runs on two instruments. Allowing for calibrations, maintenance, and rack transfers this schedule accommodates a 12 h workday.

"Intelligent" fraction collection controls the number of peaks collected through varying collection parameters such as peak slope, threshold, and wavelength(s) monitored. It is important to maintain balance to allow a good recovery of reaction products while not overloading the downstream process. On average, five peaks are collected per reaction product. However, collection parameters are variable as discussed in Section 4. A representative HPLC trace is shown in Fig. 7.3 demonstrating the capture of three fractions based on peak threshold at 219 nm.

Another factor found to have a significant impact on the efficiency of steps downfield from the HPLC fractionation is the size of vessel used for peak collection. Initially, fractions were collected into 5 ml microtiter plates. In a second

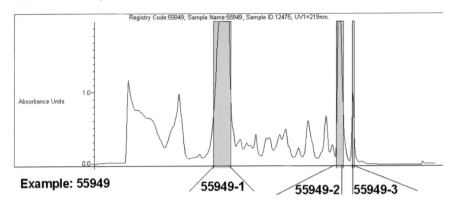

Figure 7.3   Representative preparative HPLC trace.

step, the fractions were transferred to 20 ml tared scintillation vials, combining wells when peaks spanned more than one well. This step proved to be the rate-limiting step in the overall process. The process was optimized to allow collection into 18 ml test tubes. The elimination of the liquid transfer step increased throughput from 400 fractions per day to 2000 fractions per day. Only one fraction is collected per peak to avoid combining fractions in later steps with excess volumes being sent to waste. This affects less than 10% of fractions and serves to increase the overall purity of these fractions. For smaller focused libraries, it is still possible to adapt the system to collect multiple fractions, per peak.

3.2   Identification and Prioritization

After HPLC fractions have been collected, the next step is mass spectral characterization. A 200-µl aliquot from each test tube is dispensed into a shallow well microtiter plate for analysis using a Multiprobe IIex (Packard, Downers Grove, IL) liquid handling station. Four racks of 24 test tubes are clustered to provide 96 samples in one microtiter plate.

Electrospray mass spectral data are obtained on the fractions in both positive and negative mode as discussed in Section 4; the magnitude of the evaporative light scattering detection (ELSD) signal is also captured (Fig. 7.4). The data are auto-analyzed using an Applescript™ program (PE Sciex, Ontario, Canada) that searches for peaks that correspond to the molecular ion of the product as $M + H^+$, $M + NH_4^+$, $M + Na^+$ or $M - H^-$. A Microsoft Excel™ file is produced that contains an image of each spectrum in which the peak of interest is labeled if present (Fig. 7.5). In addition, a summary is created that lists the tracking ID (TID) for each reaction product along with its collected HPLC peaks. For each TID-peak the molecular weight and molecular formula of the product are provided along with its detection status. Likewise, the molecular weights and molecular formulae of the starting materials are listed along with their detection status (Table 7.1). Finally, the molecular weight of the base peak and ELSD response are summarized. Test tubes

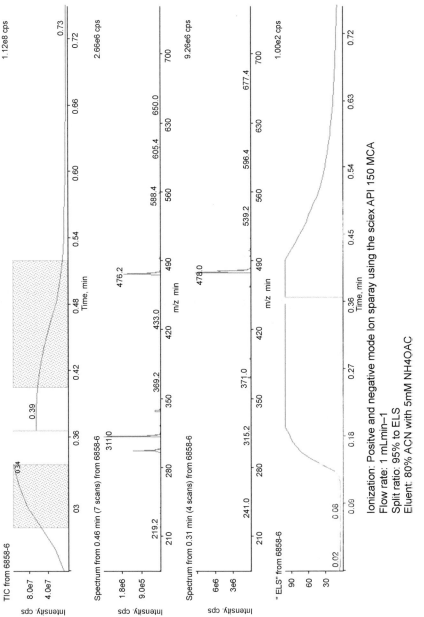

Ionization: Positive and negative mode Ion sparay using the sciex API 150 MCA
Flow rate: 1 mLmin−1
Split ratio: 95% to ELS
Eluent: 80% ACN with 5mM NH4OAC

Figure 7.4   Representative ELSD/MS data of the collected fraction.

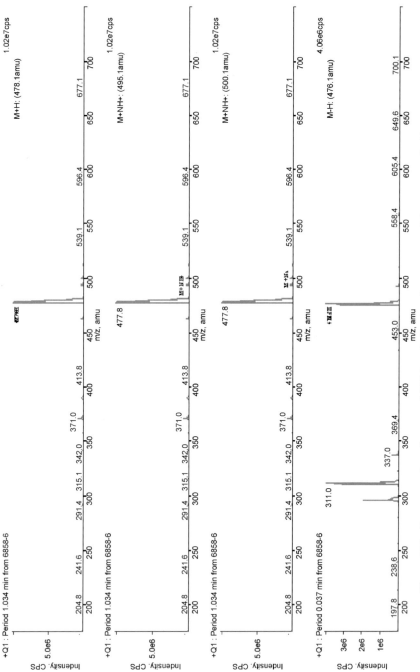

Figure 7.5 Representation of expected MS ions indicating the presence of the compound of interest.

Table 7.1   Results summary for two samples (6857 and 6858) with multiple peaks

TID-Peak	MW	Formula	Result	Base peak	ELSD area	Substrate ID	S Formula	Result	Reagent ID	R Formula	Result
6857-1	430	$C_{13} H_{11} I N_4 O_5$		307.83	958.3	S01425	$C_6H_9N_3O$		R03127	$C_7H_4INO_5$	$M - H$
6857-3	430	$C_{13} H_{11} I N_4 O_5$	$M + H$	308.03	73.4	S01425	$C_6H_9N_3O$		R03127	$C_7H_4INO_5$	$M - H$
6857-4	430	$C_{13} H_{11} I N_4 O_5$		306.83	305.9	S01425	$C_6H_9N_3O$		R03127	$C_7H_4INO_5$	$M - H$
6857-5	430	$C_{13} H_{11} I N_4 O_5$	$M - H$	722.02	20.1	S01425	$C_6H_9N_3O$		R03127	$C_7H_4INO_5$	
6857-6	430	$C_{13} H_{11} I N_4 O_5$		322.04	45.8	S01425	$C_6H_9N_3O$		R03127	$C_7H_4INO_5$	
6858-1	477.2	$C_{26} H_{24} F N_3 O_3 S$		296.03	391.5	S01425	$C_6H_9N_3O$		R01223	$C_{20}H_{17}FO_3S$	$M + H$
6858-3	477.2	$C_{26} H_{24} F N_3 O_3 S$		296.03	1018.6	S01425	$C_6H_9N_3O$		R01223	$C_{20}H_{17}FO_3S$	$M + H$
6858-4	477.2	$C_{26} H_{24} F N_3 O_3 S$		512.23	1332.5	S01425	$C_6H_9N_3O$		R01223	$C_{20}H_{17}FO_3S$	
6858-5	477.2	$C_{26} H_{24} F N_3 O_3 S$		384.05	57.3	S01425	$C_6H_9N_3O$		R01223	$C_{20}H_{17}FO_3S$	
6858-6	477.2	$C_{26} H_{24} F N_3 O_3 S$	$M + H$	478.04	2004.4	S01425	$C_6H_9N_3O$		R01223	$C_{20}H_{17}FO_3S$	
6858-7	477.2	$C_{26} H_{24} F N_3 O_3 S$	$M + H$	478.24	20.4	S01425	$C_6H_9N_3O$		R01223	$C_{20}H_{17}FO_3S$	
6858-8	477.2	$C_{26} H_{24} F N_3 O_3 S$	$M + H$	748.02	9.1	S01425	$C_6H_9N_3O$		R01223	$C_{20}H_{17}FO_3S$	

that contain the designed product or a by-product are flagged to be segregated by the automatic sorting station in a subsequent step in the process. The ELSD response is reviewed when two fractions contain the designed product; the fraction with the largest response is selected to be entered into the library. The ELSD response is also reviewed for compounds that are to be saved as by-products of possible interest; only those fractions for which the response exceeds an adjustable preset threshold are entered into this category.

3.3   Sorting by Library Type

Compounds can be classified according to the following categories: designed products, unreacted starting materials, by-products, and intermediates (unique starting materials prepared for each library). After analysis of the mass spectral data, those tubes that contain the designed product are tagged to be sorted for plating. Those test tubes that contain starting materials are tagged for disposal. Tubes that contain intermediates are summarized in a report to allow the material to be recovered if desired. Test tubes that contain neither starting materials nor designed products are assumed as by-products and are saved as compounds of possible interest. Mass spectral data on this final class of compounds is stored and will be analyzed at a later time if a hit is determined.

To reduce human error and maintain high efficiency, an automated workstation was developed for the "wet sort" process (Mettler-Toledo Bohdan, Inc., Vernon Hills, IL, Fig. 7.6). The footprint of the sorting station allows 18 racks to be placed on the deck at a time with 12 source positions and six destination positions. The 12 source positions hold racks of test tubes collected from the HPLC. The six

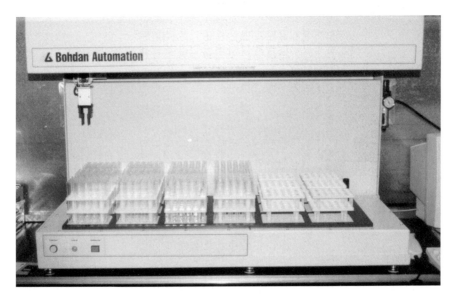

Figure 7.6   Collected fractions sorting station.

destination racks are empty. Three of the destination racks are designated for designed products and three are designated for side-products. Barcodes on source racks, containing fractions in test tubes, are scanned into the computer as they are placed on the deck. This information alerts the sort station to the compounds which are the designed products and which are the side-products. Selected test tubes are removed, by the robot, from the source racks and placed in predetermined positions in the appropriate destination racks (designed product or by-product). When all source racks or destination racks are consumed, the operator is alerted and loads in the next set of racks. The fractions contained in twelve racks of test tubes can be sorted in 30 min.

3.4 Drying and Weighing of Compounds

After sorting, it is required to dry the fractions and determine the amount of material in each test tube to control the quantity of compound plated. Drying time was a significant concern. To operate multiple drying systems would require large amounts of time for operation and maintenance. The desired system should have the capacity to efficiently evaporate large volumes of high boiling solvents from multiple samples. The centripetal evaporation system custom designed for the task was the Super-Mega Genevac (Genevac Ltd., Ipswich, England, Fig. 7.7). With a capacity of 32 racks, it can dry up to 768 samples, each containing up to 15 ml of acetonitrile/water within 8 h at 40°C. At the heart of the system is a Cole pump

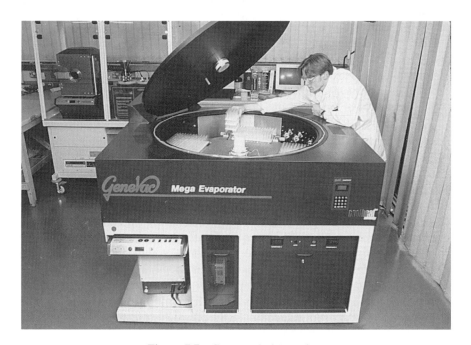

Figure 7.7   Genevac drying station.

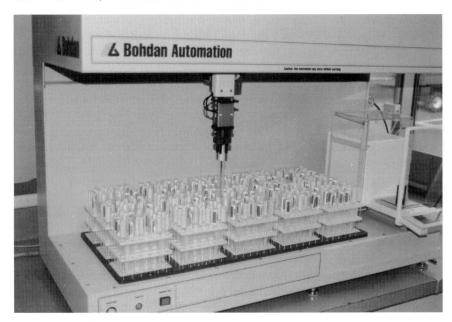

Figure 7.8   Collected fractions weighing station.

which eliminated the need for solvent condensation using a cold trap. Two solvent reservoirs connected in series allow an automatic switch to a second tank when the first is full, eliminating down time to empty tanks.

After the compounds are dry, test tubes are again taken to the weigh station (Fig. 7.8). The barcode of each test tube is scanned as it is weighed. The tare weight is then subtracted from the final weight of each test tube to determine the mass of compound in each test tube. The number of micromoles can be calculated using the molecular weight of the compound as determined by mass spectral analysis.

3.5   Sorting by Quantity

The isolated yield of each compound can vary from only a few micromoles to over 100 μmol. The goal of the process is not only to plate reaction products in known amounts but to plate the same quantity in each well of the assay plate(s). By controlling not only the purity, but also the amount of each compound, one can obtain meaningful SAR data. The number of plates that can be prepared from each purified reaction product is dictated by the amount of compound. Fractions with less than pre-defined amounts of product are sent to waste at this time.

Before plating, purified reaction products are sorted by the number of micromoles that are available in each tube. The sorting workstation is versatile and performs this "dry sort" in addition to the previously mentioned wet sort. This again increases overall efficiency by reducing both time and human error. It is desirable to have as large a number of fractions as possible for each sort as this ultimately increases the number of plates that can be prepared for each range of weights. The goal is to arrange test tubes so that the tube with the greatest amount of material is in the first position on the deck and the tube with the least amount of material is in the last position.

The deck of the sorting robot can be loaded with up to 18 full racks of test tubes containing the designed reaction products or by-products, 432 test tubes total. First, the tube in position one is manually removed and set aside. The robot then chooses the tube with the greatest amount of compound and places it in this first position. This leaves an empty spot where the second test tube resided. The computer program positionally identifies which tube should reside in this location and directs the robot to select the proper test tube to fill this spot. This process is repeated until the sort is completed. One empty position will remain. By default, the first test tube, which was set aside at the beginning of the sort, is manually placed here.

3.6   Plating of Compounds

Because each rack contains 24 tubes, three racks must be grouped together to provide the 72 compounds required to prepare one assay plate. The three racks with the greatest quantities of material are grouped together and so on. For each group of racks, the number of assay plates that can be prepared is dictated by the test tube with the lowest quantity of compound for that group of racks. Assay plates are prepared in 1 and 5 μM quantities. Liquid remaining in the bottom of each test tube is placed in a plate that will be dried, sealed and placed in archival storage at $-20°C$.

Plating is done on a Packard Multiprobe IIex (Packard, Downers Grove, IL) liquid handling station. The dried material in each test tube is diluted with 1:1 methanol:methylene chloride to bring the contents of each tube to a standard concentration and vortexed for 30 min. The solution is then removed in standard aliquots of 1× or 5× and dispensed into the assay plates.

3.7   Purity and Process Check

The final step in the process is to check the purity of a representative subset of compounds to verify the integrity of the overall process. To analyze each reaction product would be labor intensive, therefore, only 10% of the samples are checked to ensure no errors occurred in the process and to verify the purity of the final products. Selected compounds are those that reside on the diagonal of each archival microtiter plate.

The method of choice is LC/MS analysis (Fig. 7.9) due to the simplicity of the method and subsequent data analysis. The HPLC detection is provided by total ion count (TIC), ultra-violet and ELSD. The TIC allows the identification of detected products. The UV trace (254 nm) is a standard method of detection and is useful to allow comparison if additional follow-up is required. ELSD is sensitive, requires no UV chromophore, and provides a response that is proportional to the mass of compound present. Auto-analysis of the data under the different detection methods,

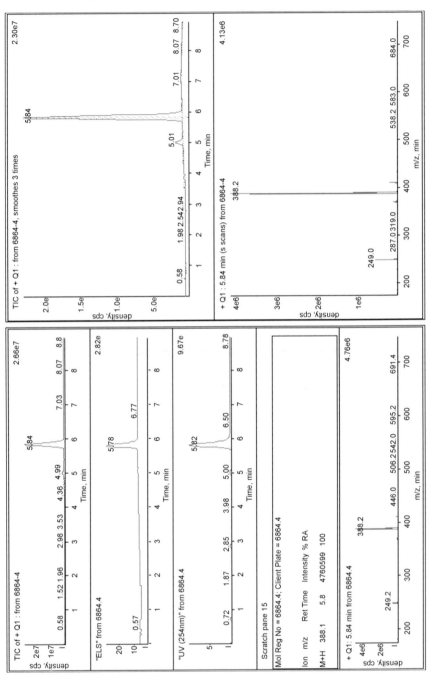

Figure 7.9 Representative LC/MS analytical data for collected fraction.

Table 7.2 Partial compilation of LC/MS purity analysis data

Well	Sample name	File name	Mass (amu)	Ret. time (min)	XIC intensity	XIC area	XIC purity	ELS intensity	ELS area	ELS purity
B3	6857-3	6857-3	430	2.44	3.90×10^5	2735967.5	9.5	1.5	19.7	30.6
A10	6858-6	6858-6	477.2	4.92	2.80×10^7	223975328	89.1	94	1082.3	96
E2	6868-3	6863-3	376.1	4.99	2.40×10^7	185389056	91.2	93	779.9	98.5
D1	6864-4	6864-4	387.1	5.84	1.60×10^7	106066232	87.69	21	156	83.2
D8	6865-4	6865-4	395.2	4.26	2.40×10^6	17171152	69.6	22	212.2	97.1
G2	6868-5	6868-5	365.2	5.82	4.80×10^7	411794656	95.8	94	1141.1	95.5
F1	6871-4	6871-4	559.2	6.07	3.40×10^7	303309920	91.8	93	1426.5	98.3
A4	6873-6	6873-6	316.1	3.27	2.10×10^6	14662063	48.2	23	199.7	95.1
C3	6874-6	6874-6	391.1	4.51	3.40×10^7	304239392	83.9	96	926.7	97
B4	6875-3	6875-3	416.2	4.46	1.50×10^7	113426288	79.6	94	966.7	98
C8	6876-7	6876-4	297.1	3.36	3.60×10^7	319951424	96.1	96	1263.8	99.5
E7	6882-1	6882-1	581.2	4.98	9.50×10^5	9420115	56.8	54	664.2	99
F9	6883-2	6883-2	461.2	7.28	1.60×10^7	104279960	97.4	94	708.8	97.6
G11	6885-3	6885-3	373.2	6.68	9.60×10^6	107497696	75.3	94	1096.4	86.2
H1	6885-5	6885-5	373.2	6.51	4.40×10^6	29086558	86	55	402.6	94.3
B9	6889-3	6889-3	351.2	5.49	2.30×10^7	161620352	98.1	95	879.3	99.8
B6	6892-2	6892-2	477.3	6.65	3.20×10^7	232074976	98	94	1516.1	96.4
C2	6894-2	6894-2	427.2	6.92	1.60×10^7	118156776	98.9	93	1095.8	98
D4	6896-2	6896-2	461.2	5.84	1.50×10^7	108215584	98.3	93	1699.3	98.1
E4	6897-7	6897-7	430.2	7.07	3.30×10^7	265033424	98.4	93	1162.8	97.5
F3	6898-1	6898-1	451.2	7.09	1.30×10^7	87235344	96	94	979	97.4
G8	6900-5	6900-5	503.2	7.01	9.60×10^6	64140208	97.8	94	1446.4	98
H10	6902-1	6902-1	423.2	5.34	1.30×10^7	87125232	97.7	94	1302.3	98
F4	6903-1	6903-1	419.2	6.55	2.10×10^7	162316992	96.9	93	1221.5	96.6

Ave. purity 93.57
Median 97.45
Std dev. 13.66

ELSD Purity Assessment

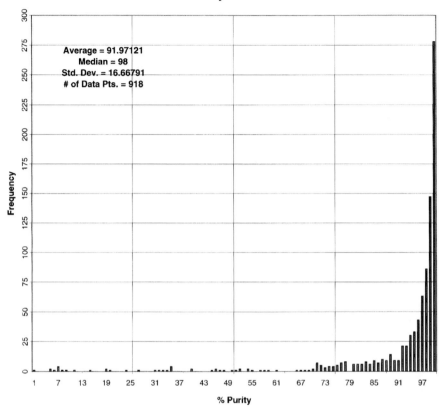

Figure 7.10   ELSD purity information for collected fractions.

using Applescript™, allows for the rapid evaluation and compilation of the LC/MS purity check (Table 7.2). A bargraph of the purity distribution of 9180 samples, by ELS detection, is shown in Fig. 7.10.

3.8 Compound Tracking System (CTS)

In our process, two Parallex HPLCs operate simultaneously to fractionate 464 samples per day producing approximately 2320 fractions. Compounds that are entered into a library will visit eight different workstations before the final plating. To track this large volume of sample information, a custom data tracking system was designed and implemented (Fig. 7.11). In addition to tracking the samples through the process, the CTS plays a key role in the identification and prioritization of fractions containing designed reaction products and by-products.

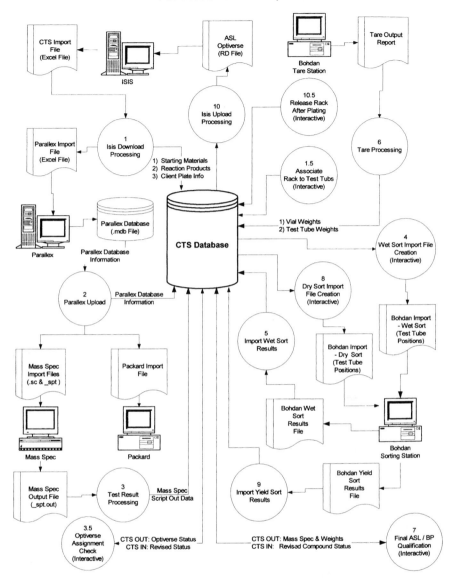

Figure 7.11   Schematic of CTS data flow.

As each reaction product is synthesized, information including the structure of the designed product, registry number, starting materials, reaction conditions, and microtiter plate barcode/well location is stored in a permanent ISIS™ database. Information required for samples to proceed through the process is downloaded to the CTS. The CTS is a temporary database. Information downloaded to the CTS includes registry number, molecular weight and molecular formula of the designed product, reaction ID, starting material IDs and starting material molecular weights, microtiter plate barcode and well location. Each sample is assigned a temporary TID to allow compounds other than the designed product to be processed and entered into a second by-product collection.

An import file is prepared for reaction mixtures that are to be submitted for HPLC fractionation containing TIDs, microtiter plate barcodes and well locations. These files are accepted by the Parallex™ workstation in Microsoft Excel™ format. The HPLC peaks are collected into pre-weighed test tubes that are held in 24 position racks. Information for the barcode on each rack is contained in the CTS such that it is associated with the test tube barcodes, test tube weights and rack location of the test tubes it holds. At the completion of each HPLC run, approximately 150 samples, each test tube holds one fraction per peak. The Parallex™ HPLC has an internal database that tracks the rack and tube into which each TID-peak is collected along with the volume of eluent, the retention time, and solvent composition associated with each peak. Once the run is complete, the Parallex™ data are uploaded into the CTS database. Using plate ID and well location, the CTS correlates the barcode on each test tube to the TID-peak and related information. The HPLC traces are converted to a Microsoft Excel™ format and uploaded to the CTS.

The next step is the preparation of sample plates for mass spectral analysis. An aliquot from each fraction is transferred to an individual well of a 96-well microtiter plate. An input file is generated by the CTS mapping the position on the deck of the robot for each rack of test tubes and microtiter plate; this allows tracking of both the starting location and final destination of each sample. The CTS also generates a sample control template for the destination plates that indicate file name (TID-peak) as well as plate and well location for the mass spectrometers.

Once a mass spectral run is complete, the data are auto-analyzed using molecular weights and formulas of the designed products and starting materials contained in the CTS database. The AppleScript™ auto-analysis produces a summary in Microsoft Excel™ format as previously discussed (Table 7.1).

The auto-analysis results are uploaded to the CTS where the information is processed. Unreacted starting materials and intermediates are sent to waste or manually recovered. The designed product is flagged to be entered into the product library while the remaining test tubes are assigned to the by-product library. Test tubes in racks are taken to the sorting robot where the barcode on each rack is read into the computer as it is placed onto the deck. This information is uploaded to the CTS which collates the identity of the test tubes in each rack and the mass spectral result for the test tubes. The CTS then creates and sends a file to the sorting station. The file contains those tubes that require sorting and their assigned location in the destination racks.

After the contents of each test tube has been dried, racks with the test tubes are again taken to the weigh station. The barcode number on each test tube and final weight is uploaded to the CTS. The CTS accesses the tare weight for each test tube and subtracts this number from the final weight to obtain the quantity of compound in milligrams for each fraction. The molecular weight of the compound is then used to calculate the final yield of compound in micromoles.

The next step in the process is to sort test tubes by quantity. As each rack is placed on the deck of the sorting robot, its barcode is scanned into the computer. This information is then uploaded to the CTS, which has a record of the tubes in each rack. A list of test tubes in order of descending quantity is then generated and downloaded to the sorting robot along with their destination position in the racks. This allows the workstation to perform a yield based sort, which places each tube in the correct rack and location.

The final step in the process is the plating of compounds. Three racks of 24 test tubes are clustered and plated into 72 wells of a 96-well assay plate. The CTS creates an input file for the Packard plating robot detailing the plate and well location for each sample; both starting location and final destination. Each plate is taken to a standard dilution and dispensed in 1 and 5 μmol quantities with the exact number of plates prepared being dependent upon the quantity of compound plated. Data on the TID-peak, compound structure (for designed product), and assay plate and well location is uploaded to the ISIS™ database.

The CTS user-interface allows for the constant tracking of reaction products, fractions and ultimately product screening plates throughout the described process. In addition to individual compound tracking, reports can be generated, reviewing processes throughout purification and identification.

4   EXPERIMENTAL METHODS

4.1   High Throughput Preparative HPLC

After the design and synthesis of the library has been completed, approximately 100 mg of each reaction mixture is placed directly into an individual well of a 48-well 5 ml microtiter plate and evaporated to dryness. Samples are dissolved in 1.0 ml of 45:45:10, methanol:DMSO:water. Examples of other solvents which could be used in place of the DMSO:methanol mixture are DMF or trifluoroethanol. The addition of a small amount of water was found necessary to dampen the solvent effect during HPLC. On average, 10% of all samples do not dissolve. These solids are separated from the mother liquors and characterized by LC/MS. Samples which are the designed products and are of 85% or better purity are entered directly into the library. The mother liquors of samples which are less than 85% pure are subjected to high throughput preparative purification as per usual.

HPLC solvents used are acetonitrile and water with 0.1% TFA buffer. Because the acetonitrile is purchased in sealed reusable 200L canisters; the buffer is added to the aqueous phase only. The columns used are 20 mm × 100 mm reverse phase with

C_{18} packing. The gradient is sharp to allow high throughput. Peaks can be collected during the entire HPLC run.

Gradient:

Step	Time (min)	Percent water	Percent ACN	Flow rate (ml/min)
Inject	0	90	10	30
Gradient	5.5	0	100	30
Wash	7.0	0	100	35
Gradient	7.1	90	10	35
Equilibrate	9.0	90	10	35

The HPLC trace is monitored at two wavelengths. The choice of wavelengths is controlled through the use of appropriate filters: 219, 254, 270, or 307 nm. Longer wavelengths are less sensitive and are more useful for highly conjugated molecules or large sample sizes. Compounds with no aromatic moieties, which have amide bonds, require detection at 219 nm. Based on design, compounds with no aromatic moieties are clustered before fractionation. Collection parameters are based on either threshold or slope/slope enable. The latter conditions are useful to avoid collection of drifting baseline by requiring that peaks exceed a preset slope enable (typically 0.5–1.0 AU) and have a rise faster than a preset slope (typically 0.05–0.1 AU per second). Parameters are adjusted to allow the collection, on average, of five peaks per fractionation.

4.2 Flow-Inject Mass Spectrometry

The identification of compounds in collected peaks is performed by flow-inject analysis using an API150 mass spectrometer (Applied Biosystems/MDS Sciex, Foster City, CA) in both positive and negative mode. Six scans are collected in positive mode (0.18 min) before the polarities are reversed to acquire data for the remainder of time in negative mode (0.58 min). Samples are injected onto the mass spectrometer using a Gilson 215 Liquid Handler (Gilson, Inc., Middleton, WI). The solvents used are 80:20 acetonitrile:water with 5 mM ammonium acetate buffer. Solvents are delivered using two HP1100 series binary pumps at a flow rate of 1 ml/min with the flow being split such that 95% goes to a Sedex 55 evaporative light scattering detector (S.E.D.E.R.E., Richard Scientific, Novato, CA) and 5% goes to the mass spectrometer. The temperature of the ELSD is set at 40°C.

4.3 Liquid Chromatography/Mass Spectrometry (LC/MS)

High-performance liquid chromatography is performed using a Monitor, C_8 reverse phase, 4.6 mm × 30 mm column. The eluent is split between the mass spectrometer and UV/ELSD in a ratio of 5/95. The organic and aqueous phase each contain 5 mM ammonium acetate buffer to facilitate ionization of the compounds. A sample size of 1 mg in 10 μl is injected. The gradient is shown below.

Gradient:

Time (min)	Percent water	Percent acetonitrile	Flow rate (ml/min)
0	90	10	1.5
5	10	90	1.5
6.5	10	90	1.5
7	90	10	1.5
7.5	90	10	1.5

5 FUTURE TRENDS

5.1 Expansion of the Decision Tree

Many side-products are the result of common side reactions and can be predicted to occur according to their reaction type. Examples of these are bis-adducts of sulfonyl chorides to amines, bis-adducts in the reductive amination of aldehydes or ketones to primary amines and hydrolysis of esters or acetates. These side-products can be as useful as the designed products in providing leads for drug candidates and could be welcome additions to a screening library. Just as algorithms can be written to predict the designed products of reactions, they can be written to predict common side-products. An example is the carboxylic acid provided by hydrolysis of a methyl ester. In addition to searching for the $M + H^+$ the program can search for the presence of the $M - 14 + H^+$ as well. Specific algorithms can be written for specific side reactions based on reaction type.

5.2 Collection of Isomers

The assumption is made that when two fractions from an HPLC separation have the same molecular weight the compounds must be isomers. This is often, but not always, true. For example, if the peak shape is not ideal, collection parameters may cause the same compound to be collected as two separate peaks. The present system allows only the major product peak per fractionation to continue through the process. Future versions of software could be written to look for the possibility of diastereomers and geometrical isomers by reviewing the compound's structure for more than one chiral center or olefins that could indicate *E/Z* isomers. If the possibility of isomers is verified, both compounds could be tagged by retention time or solvent composition and be allowed to proceed through the system. If desired,

LC/MS data could provide the retention time of each peak to further verify the two compounds have different retention times.

6 CONCLUSION

This method has proven to be versatile. By using customized automation and a dedicated CTS, targeted combinatorial libraries of 100 reaction mixtures to screening libraries of many thousands of reaction mixtures have been purified and characterized. By plating compounds in high purity and in known standard quantities, meaningful SAR data can be obtained for hits and related compounds. The utility of this method has been further increased by using it for the isolation, characterization and plating of both designed products and by-products.

A few years ago, the bottleneck was providing compounds in sufficient numbers to satisfy the appetites of high throughput screening groups. With the advent of combinatorial chemistry, the bottleneck moved downfield to the isolation and identification of reaction products from reaction mixtures in sufficient quantities to allow further follow-up. Our method has not eliminated the need for this latter step. However, by providing reaction products of high purity and known amount, purified libraries can be expected to significantly reduce the time spent on following up hits from combinatorial libraries, which will allow scientists to focus efforts in more interesting and productive areas.

ACKNOWLEDGEMENTS

The technology described herein is the result of an intense combined effort amongst synthetic, combinatorial and analytical chemists including: Tim Albertson, Dr Vincent Antle, Julie Bukowski, Dr Zane Jia Chen, Scott Clary, Cheryl DeGolier, John Dolan, Chris Doles, Dr Paul Fleming, Viktoriya Gorkina, Dr Scott Harris, Seth Harris, Marianne Hathaway, Susie Hawkey, David Heinrich, Devin Hendricks, Pat Higby, T.J. Higley, Dr Steve Hillard, Polina Kazavchinskaya, Jacob Kennedy, Mike Kobel, Cory Kolm, Halle Kunst, Solveig LaTurner, Scott McKibbin, Caroline McNaughton, Ryan Mercer, Leroy Ohlsen, Heather Ruud, Barbara Tennis, Paul Tittle, Kelly Walker, Rachel Weiner, and Wendy Wright. AXC Interactive Consultant, David R. Schedin, was critical in the development and programming of the CTS, as well as providing technical assistance and advice throughout.

REFERENCES

1.   C.D. Garr, L.M. Cameron, D.R. Schedin, and L.M. Schultz, US Patent 5,993,662 (1999).

8

Screening Single-Bead Combinatorial Libraries using Capillary HPLC and MALDI-TOF-MS

Dietrich A. Volmer

National Research Council, Halifax, NS, Canada

Oliver Keil

Graffinity Pharmaceutical Design, Heidelberg, Germany

Tammy LeRiche

University of Ottawa, Ottawa, ON, Canada

CONTENTS

1 INTRODUCTION

This chapter describes the analytical techniques developed for a high throughput synthesis and screening platform for hit optimization. The aim of the project is to integrate miniaturized chemical synthesis on single beads with biological and

High Throughput Analysis for Early Drug Discovery
Edited by James N. Kyranos

chemical screening in the 384- and 1536-well plate formats. Starting with a series of hits from an high-throughput screening (HTS) run, compound libraries are synthesized by means of solid-phase synthesis. The amount of compound subsequently released from a single polymer bead ranges between only 5 and 50 nmol. Consequently, a study was undertaken to develop a sensitive analytical tool that would allow for the identification and purity assessment of these library components using a very small fraction (typically 10% or less) of the total product released from a single bead.

Mass spectrometry is ideally suited for the analytical characterization of large combinatorial libraries because of its speed and high sensitivity.[1–3] Most analytical methods utilize high-performance liquid chromatographic (HPLC) separation combined with atmospheric pressure ionization (API) techniques such as electrospray ionization (ESI) or atmospheric pressure chemical ionization (APCI).[4–7] Matrix-assisted laser desorption ionization (MALDI) has also been used for the direct analysis of library components on synthesis beads and from micro-titer plate wells. For example, Egner et al. analyzed[8] combinatorial libraries directly from prepared polystyrene beads. Yu and coworkers used MALDI-TOF-MS for the identification of library components and side products of a single-bead cyclic peptidomimetic library.[9] Because of the matrix interferences, studies using MALDI mass spectrometry are usually focused on libraries containing components with masses $m/z > 500$.[10,11] We have recently demonstrated in our laboratory,[12] however, that MALDI-TOF-MS can be applied to the analysis of low molecular weight drug molecules in the mass range m/z 150–500 using post-source decay (PSD) and collision induced dissociation (CID). Diverse product ions for small molecules that were highly indicative of the analyte structure were obtained. Such spectra permit the fingerprinting and straightforward identification of library components in combinatorial syntheses as well as confirming the molecular weights. Furthermore, MALDI offers the advantage of a relatively high tolerance of salt in the sample.[13] This is particularly important since HPLC methods containing a phosphate buffer mobile phase are often required during various stages of the drug discovery process.[14] Standard LC/MS methods using ESI or APCI cannot readily be used under these conditions, particularly not in high throughput and automated applications such as library screening where thousands of samples are run unattended.

The automated approach utilized in this study[15] is a combination of capillary liquid chromatography, UV detection, and MALDI-TOF mass spectrometry. After synthesis and cleavage of the individual library compounds from single beads and transfer to 384-well plates, a small fraction (100 nl) of each library component is injected and separated on a 300 μm I.D. capillary HPLC column. The hyphenation with MALDI-TOF-MS is achieved by means of a micro-fraction collector and an automated peak detection system so that the peaks are collected directly onto the MALDI targets. MALDI-MS was primarily utilized because of the occasional necessity for using in-volatile chromatography buffer systems such as phosphate buffer. MALDI-MS, contrary to on-line ESI-MS methods, offers a high tolerance to salt in sample. This particular advantage becomes important in a regulated environment (GMP) where SOP-predetermined conditions containing in-volatile buffers cannot be easily changed to more ESI-MS compatible buffer systems. A second advantage of MALDI-MS is the possibility for post-HPLC run characterizations of synthesis components and by-products. This is particularly important for the structural identification of unknowns. The isolated fractions always allow going back to a particular product and performing further MALDI-TOF-MS experiments, especially in conjunction with PSD and CID. Finally, MALDI-TOF-MS offers the high throughput capability needed in this study. For method optimization, nine compounds were chosen as test library components. This group of compounds spans a wide range of polarities and has a molecular weight range of 159–372 g/mol (Table 8.1).

1.1 HPLC Separations

Because of the wide span of polarities of the library components, we employed a generic HPLC method using a gradient program with a wide range of organic modifier on a C-18 capillary separation column. The separation times for most of the single-bead library components were in the range of 5–15 min (Table 8.1).

Table 8.1   The test library used for method optimization

No.	Library component	t_r (min)	M_n
1	3-Amino-5-phenylpyrazole	3.52	159
2	5-Amino-1-phenyl-4-pyrazolecarboximide	4.25	202
3	3-Amino-1-phenyl-2-pyrazolin-5-one	5.09	175
4	5-(3-Amino-5-oxo-2-pyrazolin-1-yl)-2-phenoxybenzenesulfonic acid	6.11	347
5	Ethyl-5-amino-1-phenyl-4-pyrazolecarboxylate	9.39	231
6	2-(4-Aminophenyl)-6-methylbenzothiazole	10.28	240
7	2-Hydroxyphenyl-1-phenyl-1H-pyrazole-4-yl-ketone	11.97	264
8	3,6-Dihydro-3,6-diphenyl-1,2,4,5-tetrazine	12.71	236
9	2-Amino-4-p-tetradecylphenyl-thiazole	13.66	372

The precisions obtained when injecting 100 nl of sample were typically $<1.5\%$ RSD for the polar compounds **1–4** of the test library and $<5\%$ for the less polar compounds **5–9**. The observed retention time variations ranged between 1.5 and 2.5% RSD for components **1–4** and only *ca.* 0.2% for **5–9**.

The detection limits ranged between 100 fmol and 1 pmol for all investigated compounds (UV detection at 254 nm; Fig. 8.1 illustrates an example for a chromatogram obtained at low concentration levels). Utilizing an extended light path UV detector cell with slightly broadened peaks can further lower these numbers. For purity assessments, however, the detection limits obtained with the standard detector cell were sufficient to detect both the expected library components and unexpected compounds (side products, outcome of incomplete reactions or over-couplings *etc.*) at concentration levels as low as *ca.* 0.1% of the main component (see following discussion).

Interestingly, the factors crucially affecting the performance of the experimental HPLC setup turned out to be related to the hardware of the auxiliary components, and not the chemistry of the stationary and the organic phases: the types of 384-well plates, the plate sealing foils used to prevent sample contamination and evaporation, and the void volumes of the capillary connections between pump and micro-fraction collector (void volume problems are described in the following section). For example, with sample volumes of the individual library components between 10 and 20 µl, the best results were obtained with "V bottom" 384-well plates (individual well volume, *ca.* 120 µl) sealed with heat-sealing aluminum foil.

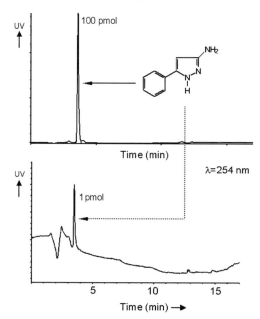

Figure 8.1 Detection limits obtained with the experimental setup used. Concentrations: 100 pmol (*upper trace*) and 1 pmol (*lower trace*).

Common PCR foils such as adhesive polymer or adhesive aluminum foils were not suited as the glue swelled after contact with the organic solvent in the samples (Fig. 8.2).

To increase sample throughput, a dual column approach was evaluated to reduce the column regeneration overhead. A second micro-pump and a column-switching device were used to equilibrate a second separation column while simultaneously performing a chromatographic separation on the first column. Unfortunately, with the short HPLC columns used, the equilibration times were quite short and no significant time savings were obtained due to the long gradient delay times of 2–5 min in capillary HPLC (dependent on flow-rate and dead volume). In experimental setups, however, where long separation columns (> 10 cm) having much longer equilibration periods are required, the potential for enhancing throughput is significant.

Figure 8.3 illustrates an example for the analysis of a single-bead combinatorial library. The figure shows a few chromatograms out of a much larger library. Unfortunately, the chemical structures of the library components are proprietary to Merck Darmstadt and cannot be identified here.

1.2   Hyphenation of MALDI and HPLC

The chromatographic peaks were collected directly as individual spots on 384-well plate MALDI targets by means of a modified commercial robotic peak collection system triggered by the UV response through an in-line detector (Fig. 8.4 illustrates the peak collection process). To avoid spillages of the mobile phase from the capillary end onto the target, a small vacuum pump removes the column effluent between the eluting chromatography peaks by means of a software-controlled moveable sheath metal capillary surrounding the fused-silica fractionation needle. When the device detects a peak by its UV response, the well plate table moves directly under the fractionation needle and the system raises the surrounding metal sheath capillary at the same time (Fig. 8.4a). Depending on the peak widths, peak volumes of 2–6 µl are collected on the MALDI target.

 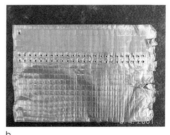

a   b

Figure 8.2   Comparison of two adhesive aluminum foils for sealing 384-well plates after contact with acetonitrile vapors for 2 days. (a) and (b) show two different manufacturers.

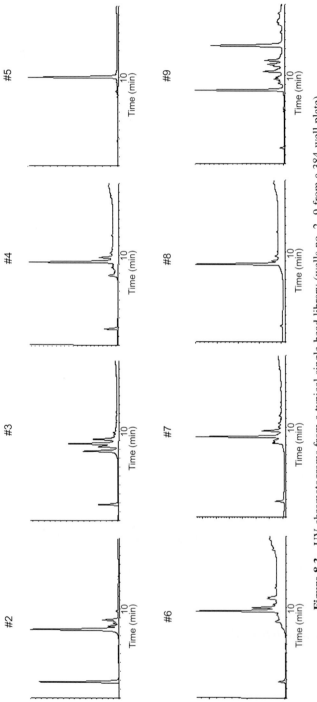

Figure 8.3 UV chromatograms from a typical single-bead library (wells no. 2–9 from a 384-well plate).

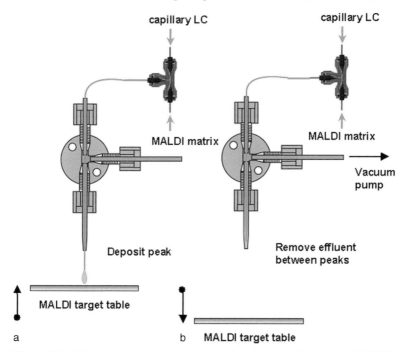

Figure 8.4   Experimental setup used for peak collection directly onto MALDI targets. (a) Peak collection from capillary HPLC onto MALDI target; (b) Effluent removal between eluting peaks.

After the entire peak is collected onto the spot, the table moves away and the sheath metal is lowered again over the fractionation needle to pump away the column effluent (Fig. 8.4b).

Important experimental parameters of the micro-fraction collection step included dead volumes, I.D. of capillaries and carry-over effects from the peak collector. Capillaries of 50 μm I.D. were used to keep the capillary volume between the detector and the fractionation needle less than 1 μl. Initially, there was concern that the peak collection apparatus (Fig. 8.4) might cause well-to-well carry-over effects. Fortunately, no significant cross talk was observed, even with slightly co-eluting peaks. For more demanding applications such as trace analysis by HPLC/MALDI, however, carry-over effects may not be negligible any longer.

1.3   Mass Spectrometric Analysis

As for all MALDI experiments, it was crucial that the right matrix was chosen for the library components. As a result, it can be expected that the matrix influences both the sensitivity and the fragmentation pattern observed. The conventional matrices of 2,5-dihydroxybenzoic acid (2,5-DHB) and α-cyano-4-hydroxycinnamic acid (α-CHCA or 4HCCA) have most commonly been used for the analysis of peptides but have also been shown to be effective for small molecule analysis.[16,17] The third matrix evaluated here, titanium dioxide (TiO_2), has recently been introduced as an alternative to the above matrices for the analysis of low molecular weight compounds due to its sparse matrix spectrum.[18] As an illustrative example,[12] Fig. 8.5 shows the fragments observed for a pharmaceutical drug molecule ($[MH-H_2O]^+$ at m/z 303) in the presence of the three matrices with both PSD and CID combined with PSD.

The fragmentation level for the MALDI matrices studied decreases in the following order: α-CHCA > TiO_2 > 2,5-DHB with both PSD and CID/PSD. In addition, α-CHCA seems to produce the most intense fragmentation level as seen from the relative intensities shown in parentheses of the signals common to all three matrices (Figs. 8.5 and 8.6). For example, m/z 232 was the most abundant fragment ion observed in the PSD spectrum using α-CHCA. In the presence of both TiO_2 and 2,5-DHB, however, the spectrum was dominated by the secondary precursor ion at m/z 303. The fragment ion at m/z 232 was observed with a relative intensity of 85% with TiO_2 and 50% using 2,5-DHB. The high degree of PSD activation achieved with α-CHCA as compared to 2,5-DHB was reported previously by Karas *et al.*[19] There, it was postulated that the strong gas-phase acidity of α-CHCA relative to other matrices such as 2,5-DHB results in the delivery of a relatively large amount of excess energy to the analyte. Therefore, subsequent metastable fragmentation of the analyte in the TOF analyzer can be expected. The organic matrices have also been classified as "hot" and "cold" according to the fragmentation level they induce. A possible explanation for this fragmentation diversity involves the temperatures at which the matrices sublime

303

PSD	PSD/CID
283	**275**
275 (45)	**258**
258 (40)	**246**
255	**232** (100)
246	**218**
232 (100)	**204**
227	**190**
199	**7**
191	
101	
10	

a

303

PSD	PSD/CID
275 (25)	**232** (40)
258 (15)	
246	
232 (85)	
206	
5	

b

303

PSD	PSD/CID
275 (10)	**232** (15)
258 (5)	
232 (50)	
3	

c

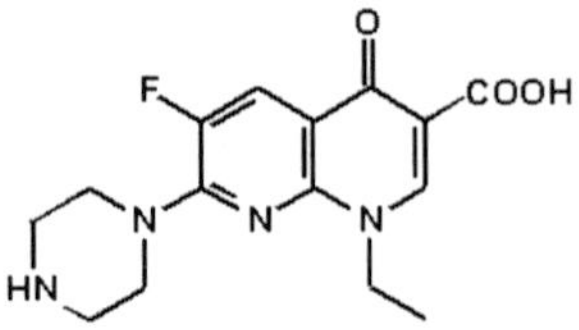

Enoxacin M_n 320

Figure 8.5 PSD and PSD/CID product ions of the $[MH^+-H_2O]$ (m/z 303) ion of enoxacin in the presence of (a) α-CHCA, (b) TiO_2, and (c) 2,5-DHB matrices. In parentheses is shown the relative intensity of the respective signal.

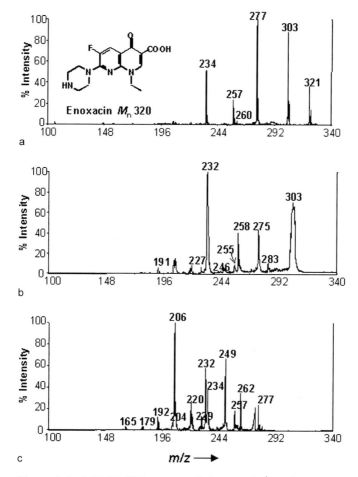

Figure 8.6 MALDI PSD spectra of the (a) MH^+ (m/z 321) ion and in-source decay product ions, (b) $[MH^+-H_2O]$ (m/z 303) and (c) $[MH^+-CO_2]$ (m/z 277) of enoxacin.

and then subsequently desorb the analytes.[20,21] On the other hand, a different mechanism involving the ionization of the analyte on the surface of the TiO_2 particles through rapid heating and vaporization has been proposed by Kinumi and coworkers[18] and Zenobi and Knochenmuss.[20]

To further investigate the influence of matrix on fragmentation and to confirm the observations gathered with small drug molecules, a structurally different substance, oleandomycin, was examined. Oleandomycin is a macrolide antibiotic, consisting of a 14-membered ring structure with two sugar residues attached. The fragments observed for oleandomycin in the presence of the three matrices with PSD and PSD/CID are shown in Fig. 8.7. As with the example in Fig. 8.6, α-CHCA was found to generate the most fragments. The differences between the quantity

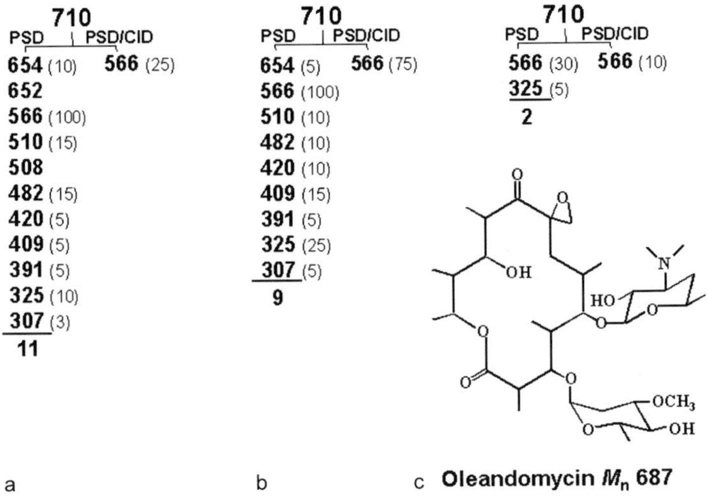

710

PSD	PSD/CID
654 (10)	566 (25)
652	
566 (100)	
510 (15)	
508	
482 (15)	
420 (5)	
409 (5)	
391 (5)	
325 (10)	
307 (3)	
11	

710

PSD	PSD/CID
654 (5)	566 (75)
566 (100)	
510 (10)	
482 (10)	
420 (10)	
409 (15)	
391 (5)	
325 (25)	
307 (5)	
9	

710

PSD	PSD/CID
566 (30)	566 (10)
325 (5)	
2	

a   b   c **Oleandomycin** M_n **687**

Figure 8.7   PSD and PSD/CID product ions of the $[M + Na]^+$ (m/z 710) ion of oleandomycin in the presence of (a) α-CHCA, (b) TiO$_2$, and (c) 2,5-DHB matrices.

and intensity of the fragments for α-CHCA and TiO$_2$, however, were not as pronounced as observed with enoxacin. 2,5-DHB was again found to be the least effective matrix of the three for fragmentation.

As molecular weight confirmation was the primary objective here, 2,5-DHB was chosen for all further investigations.

The MALDI matrix was spotted using a variety of approaches: matrix *pre-coating* of the targets before elution of the peaks , *post-coating, sandwich-spotting* and *on-line-mixing* of matrix solution and LC effluent. The matrix-to-analyte ratio was very low under these conditions in comparison to typical MALDI analyses of large biomolecules. MALDI may in fact not be the predominant ionization mechanism here, other mechanisms such as surface-assisted LDI are probably also involved in the ionization process.[22,23] (The optimum value for the matrix-to-analyte ratio was determined experimentally. When decreasing the matrix concentration down to 0.25 mg/ml, the analyte signal increased. At matrix concentrations lower than 0.25 mg/ml, however, no further increase in signal was obtained but the noise increased slightly.)

Optimum results in terms of the highest signal-to-noise and signal-to-matrix peak ratios for the investigated analytes were obtained using either pre-coating of the targets or on-line mixing of HPLC effluent and matrix solution. In the on-line mixing technique, the matrix flow (DHB at 0.25 mg/ml) was added to the column effluent at a flow-rate of 6 μl/min through a micro-tee positioned directly above the sample needle using a separate syringe pump (Fig. 8.4).

The MALDI target (stainless steel or anchor plates) had a significant influence on the quality of the mass spectral data and the sensitivity. It was found that anchor plates with a spot size of 400 μm produced excellent results. The anchor plates not only exhibited the highest sensitivity but, more importantly, also eased the determination of the sweet spots on the targets.

An example of automated peak collection and MALDI-TOF-MS analysis is shown in Fig. 8.8. The library component **6** (MW = 240) could be readily identified by its molecular ion at m/z 240. This molecular ion, however, is an unusual exception since all other investigated library components exhibited solitary quasi-molecular ions, mainly MH^+, often accompanied by $M + Na^+$ ions. More typical examples are shown in Fig. 8.9, which illustrates two mass spectra of test library components at characteristic concentration levels usually found in the single-bead libraries. (The amount of compound released from a single polymer bead typically ranged between only 5 and 100 nmol, resulting in sample amounts of 50–1000 pmol on the MALDI target after capillary HPLC separation and peak collection.)

With the analytical system, numerous single-bead libraries were analyzed. As an example, Fig. 8.10 shows a chromatogram of a synthesis reaction, where the subsequent MALDI-TOF-MS not only confirmed the molecular weight of the anticipated product but also structurally identified an unknown peak as well. Careful post-HPLC-run characterization of the peaks revealed an incomplete reaction during single-bead synthesis with the additional peak identified as one of the precursor compounds. (As mentioned earlier, the chemical structures and mass spectra of the library component are proprietary to Merck and cannot be shown here.)

2 FUTURE TRENDS

Future developments in library screening of micro- or nano-scale synthesis combinatorial libraries using MALDI mass spectrometry will include the use of parallel capillary separation columns, high-repetition MALDI lasers and parallel MALDI spotting routines to enhance sample throughput. The use of monolithic columns has shown[24] promise for ultra-fast, high-resolution separations and capillary monolithic columns will be evaluated for parallel separations. Improvements in MALDI-TOF-MS analysis of libraries will focus on automation aspects such as software techniques for automated sample recognition on the MALDI targets, new types of lasers for high-speed quantitative analysis, improved MALDI targets with enhanced sample concentration techniques (next generation Anchor chips) and on-line HPLC/MALDI interfaces for direct hyphenation with separation techniques such as capillary HPLC, capillary electro-chromatography (CEC) or capillary electrophoresis (CE).

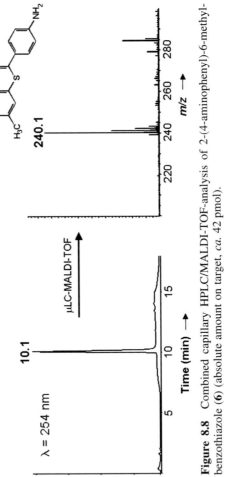

Figure 8.8 Combined capillary HPLC/MALDI-TOF-analysis of 2-(4-aminophenyl)-6-methyl-benzothiazole (**6**) (absolute amount on target, *ca.* 42 pmol).

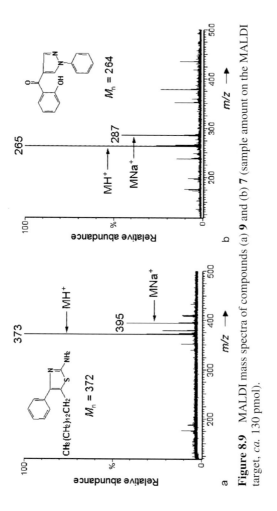

Figure 8.9   MALDI mass spectra of compounds (a) **9** and (b) **7** (sample amount on the MALDI target, *ca.* 130 pmol).

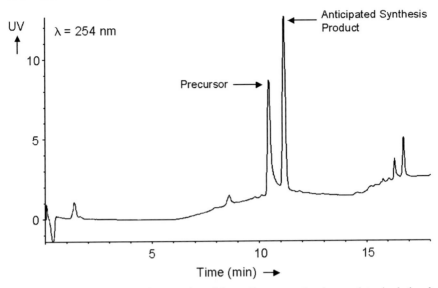

Figure 8.10   Capillary HPLC separation of the well content of an incomplete single-bead library synthesis reaction. The identities of the peaks were confirmed by the MALDI-TOF-MS analysis.

3 CONCLUSION

The analytical techniques described in this chapter allow for the rapid identification and purity assessment of single-bead nano-scale combinatorial libraries from 384-well plates using hyphenated capillary HPLC and MALDI-TOF-MS. In comparison to analytical methods using API techniques (electrospray and APCI), this methodology offers a high tolerance of salt in the mobile phase; that is, phosphate buffer-based HPLC methods can be readily used. A further advantage of MALDI-TOF-MS over on-line API methods is the possibility of performing post-HPLC separation characterization of synthesis library components. The isolated fractions always allow going back to a particular product and performing further MALDI-TOF-MS experiments, particularly in conjunction with PSD and CID, if needed. PSD and CID analyses resulted in structure informative product ion spectra. The automated interpretation and characterization of the results with appropriate software programs is currently in place as well as the extension of the technique to screening of micro-titer plates in the 1536-well plate format.

REFERENCES

1. A. Hauser-Fang and P. Voúros, *Analytical Techniques in Combinatorial Chemistry*, (M.E. Schwartz ed), Marcel Dekker, New York, pp. 29 (2000).
2. C. Enjabal, J. Martinez, and J.L. Aubagnac, *Mass Spectrom. Rev.*, **19**, 139 (2000).

3. D.G. Schmidt, P. Grosche, and G. Jung, *Rapid Commun. Mass Spectrom.*, **15**, 341 (2001).

4. Y. Dunayevskiy, P. Vorous, T. Carell, E. Wintner, and J.J. Rebek, Jr., *Anal. Chem.*, **67**, 2906 (1995).

5. T. Carell, E. Wintner, A.J. Sutherland, J.J. Rebek, Jr., Y. Dunayevskiy, and P. Vorous, *Chem. Biol.*, **3**, 171 (1995).

6. J.W. Metzger, K.-H. Wiesmüller, V. Gnau, J. Grünges, and G. Jung, *Angew. Chem. Int. Ed. Engl.*, **32**, 894 (1993).

7. J.W. Metzger, C. Kempter, K.-H. Wiesmüller, and G. Jung, *Anal. Biochem.*, **219**, 261 (1994).

8. B.J. Egner, G.J. Langley, and M. Bradley, *J. Org. Chem.*, **60**, 2652 (1995).

9. Z. Yu, X.C. Yu, and Y.-H. Chu, *Tetrahedron Lett.*, **39**, 1 (1998).

10. R. Youngquist, G.R. Fuentes, M. Lacey, and T. Keough, *J. Am. Chem. Soc.*, **117**, 3900 (1995).

11. R. Youngquist, G.R. Fuentes, M. Lacey, and T. Keough, *Rapid Commun. Mass Spectrom.*, **8**, 77 (1994).

12. T. LeRiche, J. Osterodt, and D.A. Volmer, *Rapid Commun. Mass Spectrom.*, **15**, 608 (2001).

13. O.N. Jensen, S. Kulkarni, J.V. Aldrich, and D.F. Barotsky, *Nucleic Acid Res.*, **24**, 3866 (1996).

14. A. Thomasberger, F. Engel, and K. Feige, *J. Chromatogr. A*, **854**, 13 (1999).

15. O. Keil, T. LeRiche, H. Deppe, and D.A. Volmer, *Rapid Commun. Mass Spectrom.*, **16**, 814 (2002).

16. M.-J. Kang, A. Tholey, and E. Heinzle, *Rapid Commun. Mass Spectrom.*, **14**, 1972 (2000).

17. R. Lidgard and M.W. Duncan, *Rapid Commun. Mass Spectrom.*, **9**, 128 (1995).

18. T. Kinumi, T. Saisu, M. Takayama, and H. Niwa, *J. Mass Spectrom.*, **35**, 417 (2000).

19. M. Karas, U. Bahr, K. Strupat, F. Hillenkamp, A. Tsarbopoulos, and B.N. Pramanik, *Anal. Chem.*, **67**, 675 (1995).

20. R. Zenobi and R. Knochenmuss, *Mass Spectrom. Rev.*, **17**, 337 (1998).

21. M. Karas, U. Bahr, and J.R. Stah-Zeng, *Large Ions, Their Vaporization, Detection and Structural Analysis*, (T. Baer, C.Y. Ng, and I. Powis eds), Wiley, London (1996).

22. Y.-C. Chen, J. Shia, and J. Sunner, *Rapid Commun. Mass Spectrom.*, **14**, 86 (2000).

23. Y.-C. Chen, J. Shia, and J. Sunner, *Rapid Commun. Mass Spectrom.*, **15**, 2521 (2001).

24. D.A. Volmer, S. Brombacher, and B. Whitehead, *Rapid Commun. Mass Spectrom.*, **16**, 2298 (2002).

9

The Role of NMR in the Analysis of Chemical Libraries

Samuel W. Gerritz

Bristol-Myers Squibb, 5 Research Parkway, Wallingford, CT 06492, USA

Andrea M. Sefler

GlaxoSmithKline, 5 Moore Drive P.O. Box 13398, Research Triangle Park, NC 27709, USA

CONTENTS

1 INTRODUCTION

A popular misconception within the research organizations of many pharmaceutical companies is that their primary function is to discover drugs. Strictly speaking, this is not true. Rather than physically discover a drug, pharmaceutical research organizations establish the medicinal potential of a given chemical structure

High Throughput Analysis for Early Drug Discovery
Edited by James N. Kyranos

(or structures) and transfer the appropriate supporting information to their respective development organizations. The development organization is then responsible for conducting all of the requisite activities (scale-up, formulation, stability, clinical, *etc.*) which support the submission of a new drug application (NDA). Only a small amount of the data in a typical NDA originates in the research organization. Nevertheless, the information provided by research stimulates the initiation of the development project. This information takes many forms, including biological activity, experimental conditions, synthetic routes, and, most pertinent to this chapter, analytical data for key compounds. Positive (*e.g.* the drug candidate is a potent antagonist or agonist at the receptor of interest) and negative (*e.g.* the drug candidate is inactive at a number of other receptors) data are both valuable in this effort. Not surprisingly, the information sources of greatest interest to a pharmaceutical company are those that provide intellectual property for the company to protect in a patent. In terms of revenue, the most valuable intellectual property for a pharmaceutical company is the structure activity relationship (SAR) which exists between a series of related molecules and a medicinally relevant biological system. A well-defined SAR is valuable in two ways: (1) it helps the researchers understand which molecules to make next in a chemical series in order to discover a better drug candidate; and (2) it allows a company to protect all of the active molecules in a chemical series, including many which they elect not to pursue as a drug candidate but if left unprotected would be very attractive to another company. Any technology which increases the rate and/or quality of SAR generation would thus provide a competitive advantage, and many millions of dollars have been spent by the pharmaceutical industry in pursuit of this elusive goal. High throughput screening is one such technology, and the ability to screen hundreds of thousands of samples each week has greatly accelerated the biological component of SAR generation.[1,2] Combinatorial chemistry is a technology to accelerate the chemical component of SAR generation,[3] and this chapter will focus on the need for high throughput analytical tools which ensure that combinatorial chemistry-derived samples are characterized both efficiently and sufficiently.

2   INFORMATION ISSUES IN SAR GENERATION

In the early days of drug discovery, nearly every potential drug molecule was tested in an appropriate animal model to assess its activity in a relevant biological system. This approach suffered from extremely low throughput, large compound requirements, and produced data that was confounded by many variables, including some that were species-specific. Moreover, these studies carry monetary and ethical costs that most pharmaceutical companies are anxious to avoid. Fortunately, advances in molecular biology and screening technology have provided scientists with *in vitro* assays which are the basis of the generic compound progression scheme outlined in Fig. 9.1. While differences certainly exist between companies in both the sequence and granularity of their respective compound progression schemes, they share a common overarching goal. Each step in the process is intended to generate valuable

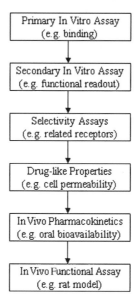

Figure 9.1   A typical compound progression scheme.

information for compounds of continuing interest to the project team, and to filter out compounds which do not have the desired properties to advance to the next step. It is often challenging for the project team to set stringent criteria for compound progression without excluding molecules that could become clinical candidates. Complicating this situation is the specter of experimental error, which can arise from a multitude of sources. Identifying and minimizing experimental error remains one of the biggest challenges facing scientists engaged in drug discovery, particularly early in the process when a single negative data point can cause a project team to lose all interest in a specific compound.

From a chemist's perspective, many sources of error in a biological assay are strictly biological – there is nothing a chemist can do to prevent them. Biologists are adept at dealing with these errors, and utilize a variety of experimental and statistical techniques to both manage and estimate the error inherent to a specific biological assay. Sample distribution is another common source of error that is not strictly chemical in origin. Distribution errors range from the misidentification of samples to variations in the amount of sample presented to the assay, and can be minimized by the establishment of careful sample tracking and distribution protocols. However, neither a reproducible biological assay nor a robust sample distribution system can overcome a source of error that is strictly chemical in origin, namely, the quantity and purity of a sample at the time of submission. Medicinal chemists are keenly aware of this source of error, and as a result the historical procedure for sample submission within most companies involves rigorous characterization of each new sample prior to distribution to assays for

Figure 9.2 Traditional roles in the synthesis, analysis and purification of a sample for screening.

screening. As shown in Fig. 9.2, the synthesis and submission of a new sample *via* traditional means is a straightforward process, with the synthesizing chemist assuming responsibility for all routine characterization and purification tasks prior to sample submission. This process does not scale well, and the widespread adoption of combinatorial chemistry for rapid sample generation has highlighted the need for improved sample characterization and submission procedures.

3 COMBINATORIAL CHEMISTRY AND SAR GENERATION

The term "combinatorial chemistry" is widely used but poorly defined. For the purposes of this chapter, we define combinatorial chemistry as any synthetic strategy that produces libraries comprising all possible combinations of a basis set of reagents. When pharmaceutical companies first became interested in combinatorial chemistry, there were many opinions on how this technology could best be applied to SAR generation. The only thing everyone could agree on was that combinatorial chemistry did not appear to be well suited for the very late stages of lead optimization, where target molecules were required in large quantities. As a result of this discord, samples generated using combinatorial chemistry were synthesized using multiple strategies and subsequently presented to assays in a variety of formats. Initial work in the area focused on the synthesis and screening of chemical libraries to provide additional sample diversity to the historical compound collection. Table 9.1 lists four common synthesis and screening combinations, as well as the requisite follow-up tasks to confirm biologically active samples. Entries 1 and 2 are indicative of early efforts in hit identification, which were focused on the generation of huge numbers of molecules (often $> 10\ 000$) using Furka's split-and-pool approach[4] to afford pooled samples for screening. While pooled samples were very efficient in terms of both synthetic and screening throughput,[5,6] from a sample characterization and submission perspective they were problematic. The purity or quantity of any member of a pool could not be measured, and it was often difficult to even confirm the presence of the potential members of a given pool. When active pools were identified, either a significant amount of resynthesis was required (unencoded pools, entry 1) or a significant amount of bead handling, rescreening, and subsequent decoding was required[7,8] (encoded pools, entry 2). By comparison to the traditional method of generating SAR (high throughput screening of a

Table 9.1   The effect of synthesis strategy on library follow-up activities

Entry	Synthesis strategy	Screening format	Activity follow-up	Issues
1	Unencoded beads (split-and-pool synthesis)	Pooled samples	Resynthesis of entire pool	Additivity of sample activities; inefficient follow-up
2	Encoded beads (split-and-pool synthesis)	Pooled samples	Redistribution of active pools into single bead per well format; decoding active pools	Additivity of sample activities; decoding infrastructure; variation in single bead yields
3	Unencoded discretes (split synthesis)	Discrete samples	Resynthesis	Inefficient library synthesis, analytical throughput
4	Encoded discretes (split-and-pool synthesis)	Discrete samples	Resynthesis	Analytical throughput

historical compound collection), these methods enjoyed much higher synthesis and screening throughput. This throughput advantage evaporated when active samples needed to be confirmed, as samples from the historical collection could be resupplied much more rapidly than active pools could be resynthesized. In addition, only positive screening data could be utilized in a pooled library screening process. If an active well was identified and the deconvolution and/or decoding activities afforded a confirmed structure with reproducible activity, then that information would provide a starting point for SAR generation. Unfortunately, none of the negative results was meaningful, as there was no way to ascertain that any of the putative structures had been presented to the assay. This gain in throughput at the expense of negative data may be acceptable when hundreds of confirmed library hits are identified from each screen. But in cases when no confirmed hits are identified, little to no information is generated about either the library samples or the biological target. When pooled library samples are screened in parallel with discrete samples from the historical collection, these differences in information content and follow-up activities are magnified. This comparison is inherently unfair, as the historical collection took decades to assemble while the chemical libraries were synthesized over a few years, but most combinatorial chemistry groups have arrived at the same conclusion: library samples compatible with existing screening processes have the best chance of success. As a result, the focus of combinatorial chemistry has shifted from the synthesis and screening of pools to the synthesis and screening of discrete samples.

Until the mid-1990s, the combinatorial synthesis of discrete samples was problematic for two reasons: (1) the inability to utilize a split-and-pool synthetic protocol made the synthesis of 10K discrete samples highly labor-intensive (Table 9.1, Entry 3) and (2) throughput limitations in sample analysis forced chemists to analyze only a subset of samples in each library. The introduction of IRORI's radiofrequency encoding technology[9] addressed the first issue (at least for solid-phase libraries; no similar technology exists for solution-phase libraries), and the second issue was addressed by advances in high-pressure liquid chromatography coupled with mass spectrometry (LC/MS) instrumentation which made it possible to characterize thousands of samples per week. When compared to screening pools of library samples that were minimally characterized at best, the synthesis and screening of discrete, well-characterized samples provides greatly enhanced compatibility with existing screening practices. In particular, negative screening data are relevant because each sample has been characterized to some extent. However, most library samples are still not equivalent to historical samples for two major reasons: (1) many library samples are submitted without chromatographic purification and are therefore generally less pure than historical compounds; (2) the assessment of purity and quantity for library samples is still prone to errors. The remainder of this chapter will describe our efforts at GlaxoSmithKline to develop high throughput methods for assessing the purity and quantity of combinatorial chemistry-derived discrete samples.

4 THE VALUE OF COMPOUND ANALYSIS

Compound analysis was not originally a high priority for combinatorial chemistry. Initially it was thought that enough effort could be put into the chemistry development to guarantee a high-quality product at the end of the synthesis. Especially with solid-phase chemistry, the hypothesis was that reactions could be driven to near-completion with excess reagents, and byproducts would simply be washed away. Of course, the realities turned out to be much more complex. Byproducts can stick to resin matrices and contaminate the final product. Combinations of reagents that were not tested during the development phase may lead to incomplete or side reactions during full library production. These and other issues that emerged during the library production resulted in large collections of compounds with suspect quality.

Still, although these problems are common in high throughput chemistry, there have been many compelling arguments against compound analysis for libraries. In the recent past, analytical characterization of large collections of compounds was nearly impossible, and even today it remains an expensive and time-consuming process. Some may argue that the impurities and side-products add extra diversity to a library, or that a compound is of little value and thus not worth analysis until it is a hit in a biological screen. The real cost, however, lies in repeatedly testing poor quality compounds, which can only generate poor quality data. When one considers that the primary tangible commodity produced by a research organization in the pharmaceutical industry is a compound collection and its associated biological data, it becomes imperative to ensure the quality of this product by analytical characterization of these compounds.

5 HIGH THROUGHPUT ANALYTICAL CHARACTERIZATION

Analytical characterization of organic compounds typically entails both confirming the compound's structural identity and purity and determining the quantity of material produced. The advent of routine, automated high-pressure liquid chromatography coupled with mass spectrometry (LC/MS) opened the door for analytical characterization of large libraries of compounds,[10] but unfortunately only provides information regarding the questions of identity and purity. The question of quantity is left uncertain, which, unfortunately, is key to obtaining SAR data.

Weighing is the simplest method of determining quantity. Many companies have developed ingenious solutions to make high throughput weighing of libraries possible and are currently employing these methods successfully. Serious problems exist with the gravimetric technique, as it requires that each sample must reside in a separate container, and if the sample contains any impurities, salts, or solvents, the measurements will be inaccurate. Also, accuracy can be a problem with sub-milligram quantities of material.

Evaporative light scattering detection (ELSD) and chemiluminescent nitrogen detection (CLND) are two other possible approaches for quantitation, both of which have the advantage of being easily coupled to LC/MS analysis. ELSD, however, suffers from average errors of $\pm 20\%$ even with well-chosen calibration standards,[11] and errors can be much higher during normal use.[12,13] CLND has a much higher intrinsic accuracy of quantitation, however, instruments can be difficult to maintain and poorly maintained instruments tend to give highly variable response accuracies. In addition, accurate results for both techniques are completely dependent on the quality of the chromatography. Overlapping peaks will invalidate the quantitation information of both CLND and ELSD.

Nuclear magnetic resonance (NMR) is another commonly used quantitative analytical technique. Quantitation by NMR is done by comparing assigned NMR signals from the compound of interest to a reference signal, which can be from an internal or external reference material.[14–17] The reference material may be any substance present at a known concentration with a measurable NMR signal, for example, even the residual proton component of a deuterated NMR solvent can be used. Recently, another technique has been reported in which this reference signal is generated by the electronics of the NMR hardware, eliminating the need for physical reference standards.[18] This technique appears to hold promise, although it requires monthly calibration to maintain accuracy. All NMR methods suffer from larger sample requirements than LC/MS/ELSD or LC/MS/CLND methods for accurate and timely analysis, although unlike these techniques, the sample is not consumed during the analysis. More vexing, however, is the issue that data analysis for NMR is complicated and not easily automated, which creates challenges for high throughput applications.

6   NMR AND CLND, HOW ACCURATE?

At GlaxoSmithKline, we have implemented two complementary methodologies for analysis of compound libraries, LC/MS/CLND and automated NMR. Our chosen method for NMR analysis involves the addition of an internal standard, 2,5-dimethylfuran as a reference for quantitation.[16] The advantage of this standard is its low boiling point, which facilitates removal of the reference compound after analysis. In addition, the signal for the magnetically equivalent furan protons resides in a relatively uncrowded area of the spectrum. Both methods have been tested and compared for accuracy using known amounts of commercially available drugs. Table 9.2 shows a representative subset of the compounds, their concentrations as measured by NMR and LC/MS/CLND (the samples for CLND were diluted 25-fold), and the error based on the known concentration. As can be seen from the data, all but one of the measurements were within 10%, and most were within 5%. In this case, both NMR and CLND provided quantitation measurements with sufficient accuracy to increase the level of confidence in biological data generated from these compounds. We next turned our attention to some "real world" examples, in which samples derived from combinatorial libraries were synthesized and quantified using both NMR and CLND. In the first case, NMR and CLND were used in parallel, while in the second case NMR was used

Table 9.2   Example of quantitation errors from NMR and CLND measurements

Drug	NMR (mM)	NMR error (%)	CLND (mM)	CLND error (%)
Caffeine	24.91	3.7	0.99	2.6
Primidone	21.41	− 6.5	0.83	− 9.4
Pyrimethamine	27.86	− 2.4	1.09	− 4.6
Tetracaine	17.48	− 1.5	0.71	− 0.7
Bupropion hydrochloride	17.37	1.9	0.65	− 4.6
Azathioprine	20.21	0.6	0.82	1.9
Phenylbutazone	14.62	− 6.3	0.56	− 11.1
Hydoquinine	19.8	7.6	0.73	− 1.5
Labetalol hydrochloride	28.59	1.1	1.14	0.8
Ondansetron hydrochloride	24.87	2.9	0.98	1.4
Naratriptan hydrochloride	13.31	− 0.9	0.55	2.4
Droperidol	16.96	− 0.5	0.68	− 0.8
Salmeterol base	20.41	2.6	0.78	− 2.0
Grepafloxacin hydrochloride	29.62	8.2	1.10	0.2
Oxiconazole nitrate	19.76	3.1	0.74	− 3.0
2,4-Dinitrobenzoicacid	52.49	3.9	2.06	1.7

exclusively for the development of the solid-phase synthesis, but only a few representative samples in the completed library were quantified using NMR while all of the samples were quantified by CLND.

7   CASE STUDY 1: ANALYSIS OF META-SUBSTITUTED ANILINOAMIDES

Utilizing a simple solid-phase synthetic sequence, 15 meta-substituted anilinoa-mides (**Structure 9.1**) were synthesized on Synphase crowns using R groups which spanned a range of steric and electronic effects. All 15 samples were >95% pure by either UV or CLND detection, and they were quantified using NMR with DMFu (in an NMR tube) and CLND. As shown in Table 9.2, the NMR and CLND yields correlated reasonably well but some significant differences were observed. The large percent error between the NMR and CLND values in entries 2 and 3 is noteworthy, but still within the experimental error observed in most biological assays. As shown graphically in Fig. 9.3, the correlation of CLND and NMR quantitation provides a slope of 1.0 but a poor r-squared value (0.58). Visual inspection of the scatterplot suggests that CLND over-estimates the amount of sample relative to NMR. A more detailed analysis indicated that the error does not correlate with a specific substituent or physical property (*e.g.* electron withdrawing groups). In cases where there is a large discrepancy between NMR and CLND quantitative measurements, our experience has been that NMR is the more accurate and reliable technique because it avoids the chromatography required for CLND quantitation (Table 9.3). If two nitrogen-containing components are not separated, the CLND measurement will not be accurate.

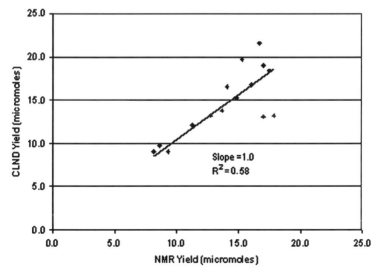

Figure 9.3   Correlation of NMR *vs* CLND yields from Table 9.3.

Table 9.3   NMR and CLND yields for 15 analogs of 1

Entry	NMR (μmol)	CLND (μmol)	CLND (% error)
1	16.0	16.8	4.8
2	12.7	13.2	3.3
3	15.3	19.7	29.1
4	16.7	21.6	29.7
5	14.8	15.2	2.7
6	11.3	12.1	6.3
7	16.9	19.0	12.1
8	17.5	18.4	5.2
9	14.0	16.6	17.9
10	8.6	9.7	12.6
11	9.3	9.0	3.2
12	8.1	9.0	10.4
13	16.9	13.1	22.7
14	13.6	13.8	1.2

8   CASE STUDY 2: ANALYSIS OF HYDRAZONE LIBRARY

A published report of hydrazones with interesting biological activities[19-21] prompted our development of a solid-phase synthesis of hydrazone analogs as shown in Scheme 9.1.[22] We performed our synthesis on Wang linker-equipped SynPhase crowns with an initial loading of 27.5 μmol per crown. The chemistry development proceeded very smoothly in terms of product purity, but we found that the average yield of final product was only 41% based on the initial crown loading. Because the purity of the final product was high, we suspected that the low yield was caused by the unintended cleavage of an intermediate from the resin rather than incomplete reaction. By quantifying each intermediate in Scheme 9.1, the source of the poor yield was identified to be the conversion of the methyl ester to the hydrazide, which resulted in the loss of approximately 30% of the resin-bound material. The release of the resin-bound material (either starting material or product) can be linked to the presence of a secondary amide bond in the "backbone" of the crown. This amide bond is required to attach the Wang linker to the aminomethylpolystyrene graft inherent to crowns, but we have observed this bond to be labile under highly basic conditions. Attempts to minimize the cleavage were largely unsuccessful, but we were satisfied that the

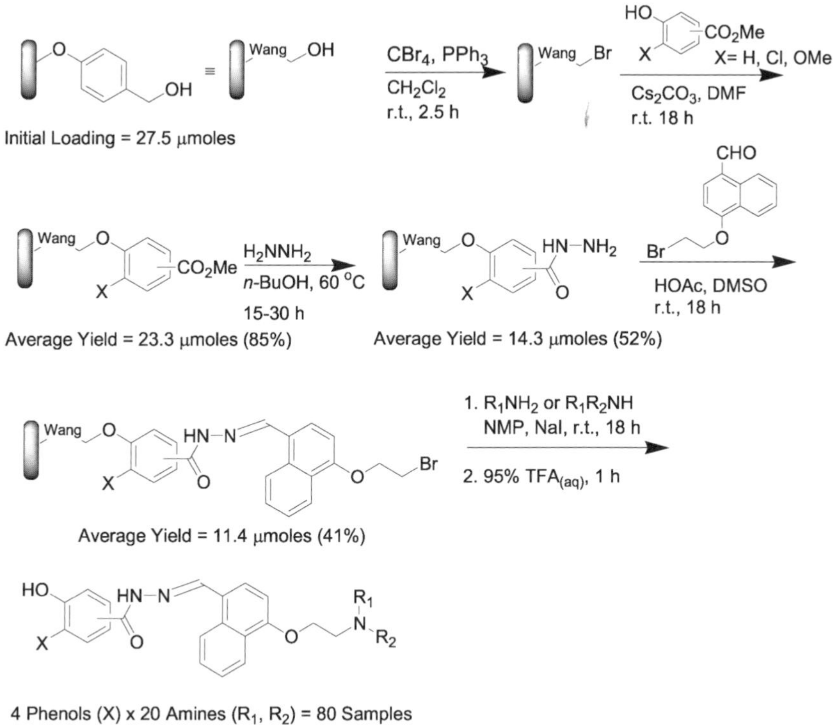

Scheme 9.1   Solid-phase synthesis of hydrazone analogs.

Table 9.4   Comparison of CLND and NMR yields for four representative hydrazones

Sample	CLND % Purity	CLND Yield (μmol)	^{1}H NMR (DMFu) Yield (μmol)	% Error
14	99	14.2	12.5	12.0
39	95	10.7	11.0	3.0
60	98	13.2	13.1	0.6
65	74	6.2	6.0	3.3

purities and yields of the final products were sufficient for screening. An 80-membered library was subsequently synthesized, and we randomly selected four samples (5% of the total) from the library and quantified the samples using both CLND and NMR (Table 9.4). As shown in Table 9.2, the CLND and NMR quantitation results correlated very well, and we elected to utilize CLND to quantitate the remaining 76 samples.

This case study illustrates the complementarity of NMR and CLND. While we were conducting chemistry development, NMR was the optimal technique because it provided both structural and quantitative information. With this information we were able to identify every proton-containing species in each sample, and thus ensured that the low observed yields were not caused by a CLND-invisible impurity but rather by a reduction in material cleaved from the resin. Note that none of the intermediates prior to the hydrazide provided a CLND signal, as they do not contain a nitrogen atom. Having utilized NMR to establish that the synthetic scheme provided samples in high purity and yield, we next determined that NMR and CLND yields were well correlated. As a result, we were able to utilize the higher throughput LC/MS/CLND technique to establish the purity and quantity of the 80 samples comprising the completed library.

9   NMR AUTOMATION TECHNOLOGY FOR SAMPLE HANDLING

Having established the value of NMR analysis in the combinatorial synthetic process, it is prudent to review the technology currently available for obtaining high throughput NMR data. The two most common methods for obtaining NMR data of large compounds arrays are: (1) robotic automation using standard NMR tubes, and (2) flow-NMR techniques using an autosampler and a flow-through NMR probe. We have worked extensively with both techniques and have found each to have its own set of advantages and disadvantages.

NMR automation using standard tubes (3 and 5 mm) is a mature technology and the simplest and most reliable type of NMR automation. This advantage exists mainly because tube-based automation has been around longer than flow-NMR techniques, which has allowed time for the development of good failure detection methods. A key strength of tube-based automation is that each sample is individually contained within its own NMR tube, eliminating sample carryover and cross-contamination issues. This strength, however, is also the primary weakness of this technique. Compounds are not typically synthesized or stored in NMR tubes, which necessitates a sample preparation step and transfer to the NMR tube. Although some manufacturers have investigated the use of modified autosamplers to prepare NMR samples, this methodology has yet to become robust and mainstream, leaving one to prepare samples manually. In addition, NMR tubes are typically too expensive to use in a disposable fashion for large numbers of samples, and the cleaning of the tubes is an issue as well.

Flow-NMR automation solves many of these problems, but introduces new ones as well. Flow-NMR uses an autosampler (typically a Gilson 215) to introduce the NMR sample into the magnet *via* tubing connected to a flow cell in the NMR probe. The autosampler can accept a large variety of sample containers, *i.e.* 96-well plates, scintillation vials, test tubes, *etc.* Thus sample preparation is simplified to dissolution of the sample in an NMR-appropriate solvent, which the autosampler can perform itself if necessary. The NMR probes are built with connections for both inlet and outlet tubing, so valves and loops can be added to create many possibilities for the sample flow path. Our group has spent a great deal of time exploring this flexibility to optimize throughput and reduce carryover.[23,24] We have found flow-NMR to be a much more complicated automation method for NMR in comparison to tube-based methods, because many fluidic issues must be considered that NMR spectroscopists are not commonly used to working with. Flow rates must be adjusted to compensate for differences in viscosity if changing solvents or concentrations. Good rinse efficiency is important to avoid carryover between samples, and, if a different solvent is used for the rinse and the sample, solvent compatibility must be considered. The solvent chosen must also be compatible with the tubing used inside and outside the NMR probe. The volumes used must be calibrated and delivered accurately and reliably to the NMR probe to ensure good lineshape in the NMR spectrum. Any air bubbles or incomplete fluid mixing inside the NMR probe will also degrade the lineshape. These issues make flow-NMR a difficult and complicated technique to implement. Another drawback to this type of automation is poor failure detection; currently leaks and clogs in the unattended system will go undetected and large numbers of samples can be ruined if the problem is not recognized and corrected promptly by the operator. Finally, thorough rinsing can consume large amount of expensive, deuterated NMR solvents, or, if protonated solvents are used, issues with solvent suppression techniques and the spectral regions they obscure must be resolved.

In general, both tube-based and flow-NMR automation is capable of handling samples from small collections (<1000) compounds. Each technique has its strengths and weaknesses and these should be considered when choosing the type of automation for a collection of samples. Clearly, NMR analysis of compound libraries would benefit from improvements to automation.

10 NMR DATA ANALYSIS

Obtaining the NMR data, however, is only part of the problem; the data must also be analyzed. NMR data are very complex and information rich, and has proved recalcitrant to thorough and completely automated interpretation. One difficulty is that the primary data, the proton NMR spectrum, is visual and qualitative; one asks the question "does this spectrum *look* right for my compound?" Compare this to the most common questions asked of LC and MS data, *i.e.* "what is the percent purity of my compound?" and "does my compound give the correct molecular weight?" These questions are very quantitative in nature and can be asked and answered by a computer relatively easily. Software programs and tools are available to aid in NMR interpretation and our group has found them useful to speed the workflow. However, we still find a degree of manual interpretation necessary. The following case study illustrates where and how we have found computer-aided analysis to be helpful for NMR data interpretation. The software our company has purchased is the suite of analytical software from Advanced Chemistry Development;[25] and the pictures used in the case study were generated using this software.

11 CASE STUDY 3: ANALYSIS OF 96 BENZOYLATED AMINES

Proton NMR data were obtained for a library of 96 benzoylated amines dissolved in d_6-DMSO containing 5 mM DMFu. The spectra were processed using ACD's CombiNMR v 5.0 software.

First we must consider which of the fundamental analytical questions (purity, identity, or quantity) we wish to assess from the NMR data. Purity is a very difficult question to answer quantitatively by NMR, but we find the NMR spectra to be useful to check the validity of the NMR obtained by UV analysis. Often impurities, particularly small amines used frequently in combinatorial chemistry, will not have a UV chromophore, leading to an overly optimistic purity measurement. Most impurities, however, do contain protons, so they will always appear in the NMR spectrum. We found it easiest to take a printed stack of NMR spectra and manually sort the spectra by the "appearance" of purity, *i.e.* good, fair, and poor. Even a large amount of spectra can be quickly sorted this way. We then cross-check the NMR purity analysis with the UV purity numbers to spot any discrepencies. Impure compounds that still appear by NMR to contain significant amounts of product can then be selected for further purification.

Confirming the identity of a compound is best conducted with a combination of multiple analytical techniques and software analysis. The ACD CombiNMR software package will predict the proton NMR spectra of a given set of structures and then compare the predicted spectra with the experimental data to attempt structural verification. The software then displays the familiar red/yellow/green light display similar to many mass spectrometry software packages. We found the automated analysis to be greatly improved by the appropriate setting of "dark regions" in the spectra, *i.e.* portions of the spectra that are ignored by the analysis such as large solvent signals. In addition the software can "learn" from example compounds, manual input of assigned spectra from just one or two analogs can vastly increase the accuracy of the spectral prediction. Figure 9.4 shows a sample display for the benzoylated amine library.

The right side of the display contains the familiar color analysis of the plate, the left side shows the calculated and experimental spectra for the selected well with the putative structure and parameters on the bottom. We have found this software analysis to be useful to focus the manual interpretation efforts towards a subset of compounds. In particular, the compounds that are marked yellow from the analysis need manual interpretation of the spectra. Mass spectral data can also be used to reduce the amount of data requiring time-consuming analysis by a spectroscopist. Particularly difficult structural questions can be answered by obtaining further NMR information such as COSY or HMQC data. Such spectra can now be obtained in 10 min or less on a 500 MHz NMR and the inclusion of 2D NMR data may greatly improve the validity of automated spectral verification in the future.

Finally, the question of quantitation can also be facilitated by the software. If a relatively isolated signal common to the majority of the compounds in the library can be selected, one can perform a quantitative analysis of the spectra using the

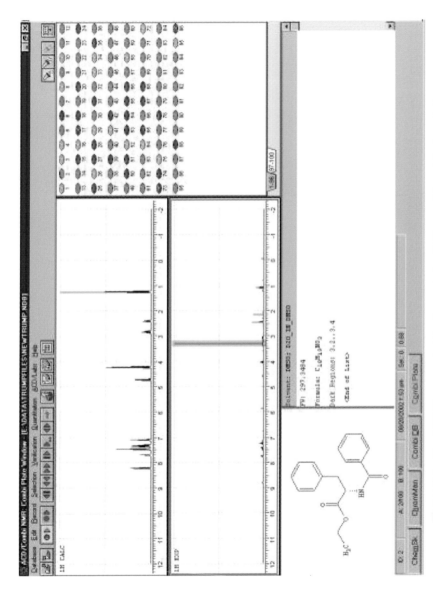

Figure 9.4   CombiNMR screen shot for the analysis of *N*-benzoylphenylalanine ethyl ester.

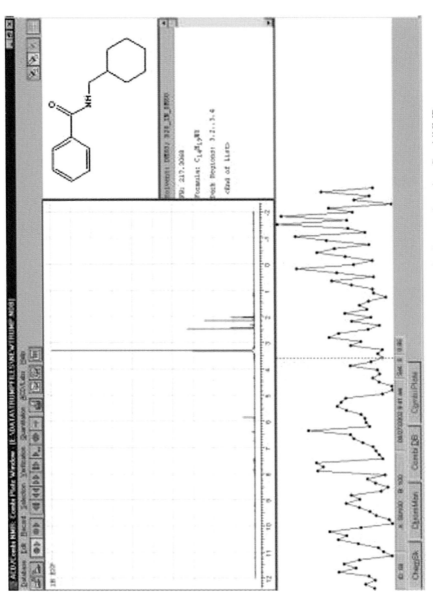

Figure 9.5   Screen shot of the automated quantitation results using CombiNMR.

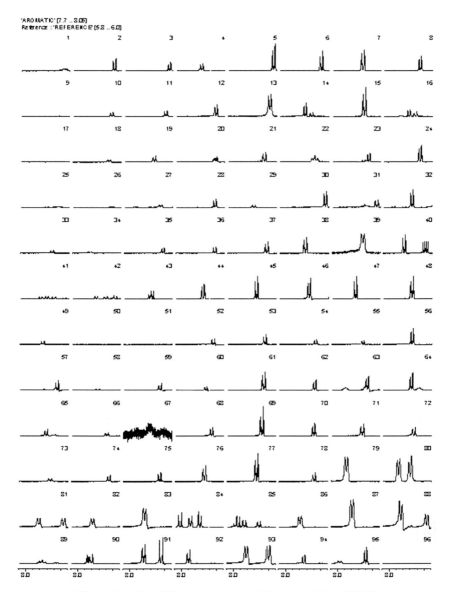

Figure 9.6   The "96 spectra at a time" feature of CombiNMR.

ACD CombiNMR software. The analysis is performed by comparing the integral of the region containing the internal standard (DMFu at ~5.8 ppm for our case) to the integral of the selected region (~7.8 ppm corresponding to a aromatic proton ortho to a benzyloxy group). Figure 9.5 shows a display of the results of the analysis for our library.

The top of the display shows the spectrum and structure of a selected well, in this case a compound obtained in poor yield (compare the aromatic signals to the DMFu signal at 5.8 ppm). On the bottom is a graphical display of the relative "yields" for all the wells; the higher the point, the higher the yield. Some of the tallest points at the end are due to more than one signal being present in the selected region, however, in general one can get a good feel for the relative yields in the different wells rapidly without individually interpreting and integrating each spectrum.

In addition, the software offers another display option to aid in the quantitative analysis of the spectrum. Figure 9.6 provides an example of this type of display for our library. Here we have chosen to display, in plate format, a selected region of the NMR spectra, from 7.7 to 8.05 ppm containing the characteristic ortho protons of the benzoyl moiety common to all the compounds in this library. The spectra, however, are all scaled to a peak from the DMFu internal standard, so the relative heights of the aromatic signals reflect the relative yields for each well. One can quickly spot the compounds obtained in good yield, *i.e.* well numbers 5,6, and 48 for example. Problems are easily spotted as well, such as poor yields (*e.g.* well 18), mixtures (*e.g.* well 16), solubility problems (*e.g.* well 67), compounds containing multiple protons in this region which give skewed results in the quantitative analysis (*e.g.* well 80) and issues with NMR performance such as poor shimming (*e.g.* well 39). This picture provides a rapid assessment of the success of a library synthesis. The quantitative results, if needed, can be obtained from integration results.

What we have learned is that the NMR data is quite valuable, and can be rapidly analyzed to obtain a rough, qualitative assessment of purity, identity and quantity. More detailed interpretation of the NMR spectra can be performed as necessary for selected compounds.

12   CONCLUSION

Our experiences with the high throughput analysis of combinatorial libraries *via* NMR and LC/MS/CLND have led us to establish a number of guiding principles, and these will be discussed in the context of the type of library effort being explored.

When developing a library-amenable synthetic route, *every single experiment* should be analyzed using NMR with an internal standard. LC/MS is useful for confirmation of the mass ion and qualitative purity measurement, but it is easy to get fooled by samples (or impurities) with large extinction coefficients. LC/MS/CLND is slightly more useful than LC/MS, but neither technique affords the structural details provided by NMR. The throughput requirements for chemistry development are generally small, and thus flow-NMR is not necessary.

In some cases, the next step in chemistry development is a reagent rehearsal, in which every potential reagent at each diversity position is tested using the optimized synthetic route. Reagents that fail to provide the expected product with high purity and yield in the rehearsal are usually removed from the reagent list prior to the production of the final library. Because the rehearsal leads to a go/no go decision for each reagent, the need for high quality analytical data is paramount. We feel that both NMR (with an internal standard) *and* LC/MS/CLND data should be collected for the rehearsal samples. In addition to supporting the selection of reagents for the production library, these data are useful for establishing a correlation between NMR- and CLND-based quantitation methods (*vide infra*).

In the case of small libraries (< 1000 discretes), a correlation must be established between NMR- and CLND-based quantitation (and in rare cases UV or ELS) in order to support the use of the higher throughput technique without compromising sample quantitation. In the case of a poor correlation, quantitation by NMR provides the most accurate results. NMR automation, either flow-NMR or tube-based NMR with automated sample preparation, can facilitate the collection of hundreds of NMR spectra in a relatively short period of time, but the analysis of these spectra is both tedious and time-consuming. An alternative approach is to collect all of the NMR data but analyze samples on an as-needed basis (*i.e.* to confirm the structure and quantity of an active sample).

In the case of large libraries (> 1000 discretes), NMR data cannot reasonably be collected on every sample, and so establishing a good correlation between NMR quantitation and any other technique is essential. In the absence of a good correlation, gravimetric analysis may be a viable option.

Even though combinatorial chemistry offers the chemist huge improvements in synthetic throughput, there is no shortcut around the rigorous analysis of each sample. This situation will not change until the chemistries underlying library syntheses are demonstrated to provide each product in 100% yield and 100% purity (even *predictable* yields and purities would be a good starting point). Even if this lofty goal is reached, samples produced *via* combinatorial chemistry beg the question "how pure is that sample?"

ACKNOWLEDGEMENTS

The authors would like to thank Ryan Trump and Robert Wiethe for their scientific contributions to this effort.

REFERENCES

1. J.P. Devlin (ed.), *High Throughput Screening: The Discovery of Bioactive Substances*, Marcel Dekker, New York, NY (1997).
2. W.P. Janzen (ed.), *High Throughput Screening: Methods and Protocols*, Humana Press, Totawa, NJ, p. 190 (2002).
3. N.K. Terrett, *Combinatorial Chemistry*, Oxford University Press, New York, NY (1998).
4. A. Furka, F. Sebestyen, M. Asgedom, and G. Dibo, *Int. J. Pept. Protein Res.*, **37**, 487 (1991).

5. X.-Y. Xiao, R. Li, H. Zhuang, B. Ewing, K. Karunaratne, J. Lillig, R. Brown, and K.C. Nicolaou, *Biotechnol. Bioeng. (Comb. Chem.)*, **71**, 44 (2000).
6. D.S. Tan and J.J. Burbaum, *Curr. Opin. Drug Discov. Dev.*, **3**, 439 (2000).
7. A.W. Czarnik, *Curr. Opin. Chem. Biol.*, **1**, 60–66 (1997).
8. C. Barnes and S. Balasubramanian, *Curr. Opin. Chem. Biol.*, **4**, 346–350 (2000).
9. K.C. Nicoloau, X.-Y. Xiao, Z. Parandoosh, A. Senyai, and M.P. Nova, *Angew. Chem. Int. Ed.*, **34**, 2289 (1995).
10. L. Zeng, X. Wang, T. Wang, and D.B. Kassel, *Comb. Chem. High Throughput Screen.*, **749**, 3–13 (1998).
11. L. Fang, M. Wan, M. Pennacchio, and J. Pan, *J. Comb. Chem.*, **2**(3), 254–257 (2000).
12. C.E. Kibbey, *Mol. Divers.*, **1**(4), 247–258 (1996).
13. I.M. Mutton, *Quantitation Without Standards. And evaluation of the Sedex 55 Evaporaative Light Scattering Detector*. Glaxo Wellcome Internal Publication (1997).
14. D.D. Traficante, *Concepts Magn. Reson.*, **4**, 153–160 (1992).
15. B.C. Hamper, S.A. Kolodziej, A.M. Scates, R.G. Smith, and E. Cortez, *J. Org. Chem.*, **63**, 708–718 (1998).
16. S.W. Gerritz and A.M. Sefler, *J. Comb. Chem.*, **2**, 39–41 (2000).
17. V. Pinciroli, R. Biancardi, N. Colombo, M. Colombo, and V. Rizzo, *J. Comb. Chem.*, **3**, 434–440 (2001).
18. S. Akoka, L. Barantin, and M. Trierwiler, *Anal. Chem.*, **71**, 2554–2557 (1999).
19. A. Ling, A. Kuki, S. Shi, M.B. Plewe, J. Feng, L.K. Truesdale, J. May, D. Kiel, P. Madsen, C. Sams, and J. Lau, *Hydroxybenzoylhydrazones of Aromatic and Heterocyclic Aldehydes as Glucagon Antagonists/inverse Agonists*, WO 0039088 (2000).
20. A. Ling, Y. Hong, J. Gonzalez, V. Gregor, A. Polinsky, A. Kuki, S. Shi, K. Teston, D. Murphy, J. Porter, D. Kiel, J. Lakis, K. Anderes, J. May, L.B. Knudsen, and J. Lau, *J. Med. Chem.*, **44**, 3141–3149 (2001).
21. A. Ling, M. Plewe, J. Gonzalez, P. Madsen, C.K. Sams, J. Lau, V. Gregor, D. Murphy, K. Teston, A. Kuki, S. Shi, L. Truesdale, D. Kiel, J. May, J. Lakis, K. Anderes, E. Iatsimirskaia, U.G. Sidelmann, L.B. Knudsen, C.L. Brand, and A. Polinsky, *Bioorg. Med. Chem. Lett.*, **12**, 663–666 (2002).
22. S.W. Gerritz, J.A. Linn, D.H. Drewry, and A.L. Handlon, *The Importance of Quantitation in Parallel Synthesis*, SRI New Chemical Technologies: Accelerating Drug Discovery Conference, San Diego, CA (2001).
23. A.M. Sefler, K.C. Lewis, and J.F. Sefler, *VAST at Glaxo Wellcome* Varian User's Group Meeting, Palo Alto, CA, April (2000).
24. A.M. Sefler, G.F. Dorsey, R.D. Rutkowske, and T.D. Spitzer, *Alternative Plumbing Schemes for Direct-Injection NMR* Experimental Nuclear Magnetic Resonance Conference, Orlando, Florida, March (2001).
25. CombiNMR v 5.0, Copyright 1994–2001 Advanced Chemistry Development Inc., All rights reserved. Since publication of this article the software has undergone substantial improvements. Version 8.0 is now commercially available.

Subject Index